SpringerBriefs in Earth Sciences

Series Editor

Pedro Maciel de Paula Garcia, Universidade Federal de Mato Grosso, Cuiabá, Mato Grosso, Brazil

W0235296

SpringerBriefs in Earth Sciences present concise summaries of cutting-edge research and practical applications in all research areas across earth sciences. It publishes peer-reviewed monographs under the editorial supervision of an international advisory board with the aim to publish 8 to 12 weeks after acceptance. Featuring compact volumes of 50 to 125 pages (approx. 20,000–70,000 words), the series covers a range of content from professional to academic such as:

- timely reports of state-of-the art analytical techniques
- bridges between new research results
- snapshots of hot and/or emerging topics
- literature reviews
- in-depth case studies

Briefs will be published as part of Springer's eBook collection, with millions of users worldwide. In addition, Briefs will be available for individual print and electronic purchase. Briefs are characterized by fast, global electronic dissemination, standard publishing contracts, easy-to-use manuscript preparation and formatting guidelines, and expedited production schedules.

Both solicited and unsolicited manuscripts are considered for publication in this series.

Shuning Dong · Hao Wang · Wanfang Zhou

Prevention and Reclamation of Mining-Induced Land Subsidence

 Springer

Shuning Dong
Xi'an Research Institute
China Coal Technology and Engineering
Group Corporation
Xi'an, China

Hao Wang
Xi'an Research Institute
China Coal Technology and Engineering
Group Corporation
Xi'an, China

Wanfang Zhou
Zeo Environmental, LLC
Knoxville, USA

ISSN 2191-5369   ISSN 2191-5377 (electronic)
SpringerBriefs in Earth Sciences
ISBN 978-3-031-78157-5   ISBN 978-3-031-78158-2 (eBook)
https://doi.org/10.1007/978-3-031-78158-2

This Springer imprint is published by the registered company Springer Nature Switzerland AG
The registered company address is: Gewerbestrasse 11, 6330 Cham, Switzerland

If disposing of this product, please recycle the paper.

Introduction

Ground subsidence is a universally recognized geohazard in response to underground coal mining. Its nature and extent are affected by many factors including mining methods, depth of extraction, thickness of extracted coal seam, and properties of the rock mass overlying and underlying the coal seam. The impacts of mining-induced subsidence are severe in terms of damage to the environment, such as deformation of land, changes in surface water and groundwater conditions, and adverse effects on ecology. Mining-induced subsidence may lead to fatalities and disasters where mining is conducted beneath populated urban and industrial areas, surface water bodies, or aquifers. In many cases, the environmental damage is irreversible, and the area is no longer inhabitable. Case studies have demonstrated that restoration of the subsided areas to their beneficial uses is costly.

Although subsidence cannot be eliminated, it can be reduced or controlled in areas where deformation of the ground surface is predicted to produce dangerous or costly effects. Concerted efforts to understand mining-induced subsidence processes in China in the last 20 years have made significant progresses in many aspects including theoretical approaches in subsidence analysis and prediction, empirical methods in subsidence prediction based on local hydrogeological conditions, engineering measures to reinforce rock layers overlying the coal seam to prevent subsidence from occurring, optimization of mining methods so that mining operations result in minimized and controlled subsidence and innovative techniques in restoring subsided areas to beneficial uses with multidisciplinary approaches.

This book addresses this important geohazard in five chapters. Chapter 1 gives an overview of subsidence processes and their harmful consequences. Chapter 1 also describes the diverse types of subsidence associated with mining and the general remedial alternative analysis for remediation. Chapter 2 describes the techniques to investigate, predict, and monitor the subsidence, in particular, application of geophysical techniques. Chapters 3–5 provide real-world case studies from China to illustrate typical engineering measures to control and reduce subsidence impacts. Chapter 3 addresses the application of an adaptive grouting approach to remediate an old mining area for residential uses. Chapter 4 introduces a new paste material and demonstrates its effectiveness in filling and grouting the underground goafs to prevent subsidence. The new paste material helps reduce the cost and improve the effectiveness of the remediation. Since groundwater dynamics are important contributors to many mining-induced subsidence, Chap. 5 describes a case study in which a water inrush incident in a coal mine is mitigated to avoid adverse impacts on ground stability and safety of mine operations. The concurrent drilling and grouting processes can be applicable to emergency rescue when mining-induced geohazards suddenly occur. The directional drilling technique also helps identify the concealed groundwater passageways and implement targeted grouting.

There is likely a conflict between increasing mining efficiency and reducing the impacts of mining on people and the environment. It has been challenging to strike a balance between high-intensity mining in large-scale modern mines and minimization of impacts on receptors of the surrounding areas. The book is intended to promote the concept of environmentally responsible mining with an objective to protect directly affected people and the environment.

Xi'an, Shaanxi, China   Shuning Dong
Xi'an, Shaanxi, China   Hao Wang
Knoxville, USA   Wanfang Zhou

Contents

Chapter 1
Mining-Induced Ground Subsidence

Abstract Underground mining created voids or mine-out areas that cause changes in the magnitude and orientation of the in-situ stress field and induce earth fractures and ground subsidence. Although some subsidence incidents are anticipated in mining areas, they can strike with little or no warning. Mine subsidence is a geohazard that can result in catastrophic and costly damage and even fatalities. Ground movement is typically swift and sudden for pit subsidence, while trough subsidence, which can develop over mines of any depth, appears as a gentle depression in the ground and can spread over an area as large as several square kilometers. Many factors including mining method, depth of extraction, size and configuration of openings, rate of advance or extraction, coal thickness, topography, lithology, structure, hydrology, in situ stresses, and rock strength and deformational properties affect the initiation and maximum lateral extent of subsidence; the most important ones are extraction width, coal thickness, and mining depth.

Keywords Underground mining · Pit subsidence · Trough subsidence · Geohazard · Extent of subsidence

Ground subsidence is a global problem and can be caused by both natural processes and human activities. The natural processes include the dissolving of soluble rocks on karst terrains, earthquake, or volcanic activity, while human activities include withdrawal of groundwater, leakage of near-surface utilities, or underground mining. This book addresses the subsidence caused by underground mining.

1.1 Type of Mining-Induced Subsidence

Underground mining created voids or mine-out areas that cause changes in the magnitude and orientation of the in-situ stress field and induce earth fractures and ground subsidence. Mining-induced subsidence is defined as the lowering of the ground surface due to movements and displacements of the overlying strata into the voids that are created by underground mining. Two types of subsidence are associated with underground mining: sinkhole subsidence and trough subsidence.

1.1.1 Pit Subsidence

Sinkhole subsidence, also known as sinkhole subsidence or chimney subsidence, occurs in areas where the mining depths are relatively shallow. This type of subsidence is fairly localized in extent and often associated with the roof collapse of mines that have total overburden of less than 50 m and weak roof rock of shale or mudstone. The term "overburden" is used here to refer to both the unconsolidated material and rock overlying the mined minerals. Such a subsidence is characterized by an abrupt collapse of the overburden into the underlying mine voids, resulting in a circular, craterlike feature that has an inward drainage pattern. In overburdens that are composed of sandy soil, the subsidence can be disk-shaped, whereas in overburdens composed of clayey soil, the subsidence is more of bell-shaped, and voids may be created in the soil prior to depression appearance on surface. Figure 1.1 shows examples of the pit subsidence in which the overburden is composed of sandy soil and clayey soil, respectively. The pit subsidence generally does not hold water; however, it can drain water directly to the underlying mine. Figure 1.2 shows the typical shape of a mine collapse.

The pit subsidence associated with room-and-pillar mining of coal or solution mining of salt and gypsum is often in this category. Because of the several stages of coal removal and the slow pillar deformation and deterioration in room-and-pillar mining, surface settlement is not uniform and immediate, rather, it may be erratic, intermittent, and long delayed. Consequently, occurrence of the sinkholes is difficult to predict. Neither the time nor the location of subsidence can be predicted with confidence in solution mining of salt because salt deforms slowly in a complex manner, and the deformations are different from those of overlying strata.

The pit subsidence can be caused by dewatering activities in mines to prevent water hazards. This is particular common in karst areas that are underlain by soluble rocks such as limestone or dolomite. Mining of coal, lead and zinc, gold, and iron ore deposits in karst areas has been closely associated with sinkholes. Many mineral deposits in China such as coal, iron, lead and zinc, gold, aluminum, and copper are located in between, or above, or below the karst aquifers. Because drainage of the karst water is essential for mine safety, they are often referred to as karst water-impregnated deposits. Most of the well-known deposits with large quantities of water (mine drainage over 1 m^3/s) are karst water-impregnated deposits, and it is in those mines that sinkholes frequently took place. Foose (1967) reported that extensive dewatering at the West Driefontein Mine near Johannesburg, South Africa, caused many sinkholes, including one in 1962 under the crushing plant that was destroyed, and 29 lives were lost. Abkemeier and Stephenson (2003) reported that a quarry in Missouri that had been dry for more than 50 years intercepted a sediment-filled cave system below the groundwater table. Dewatering of the quarry created a flow

(a)

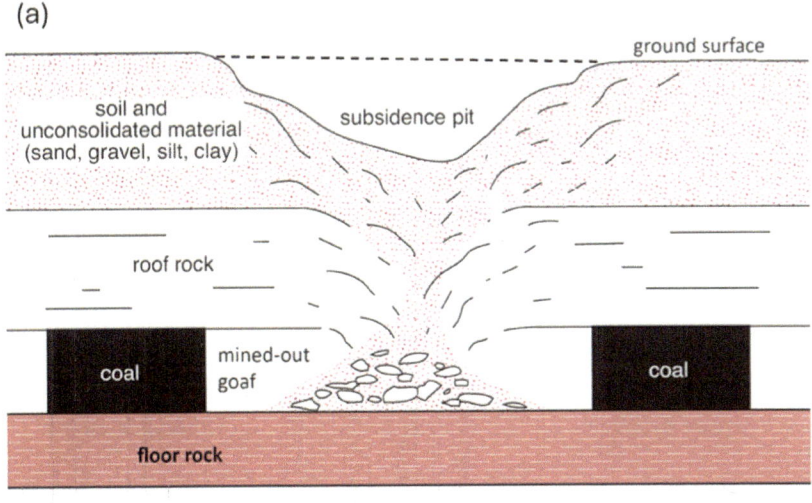

(b)

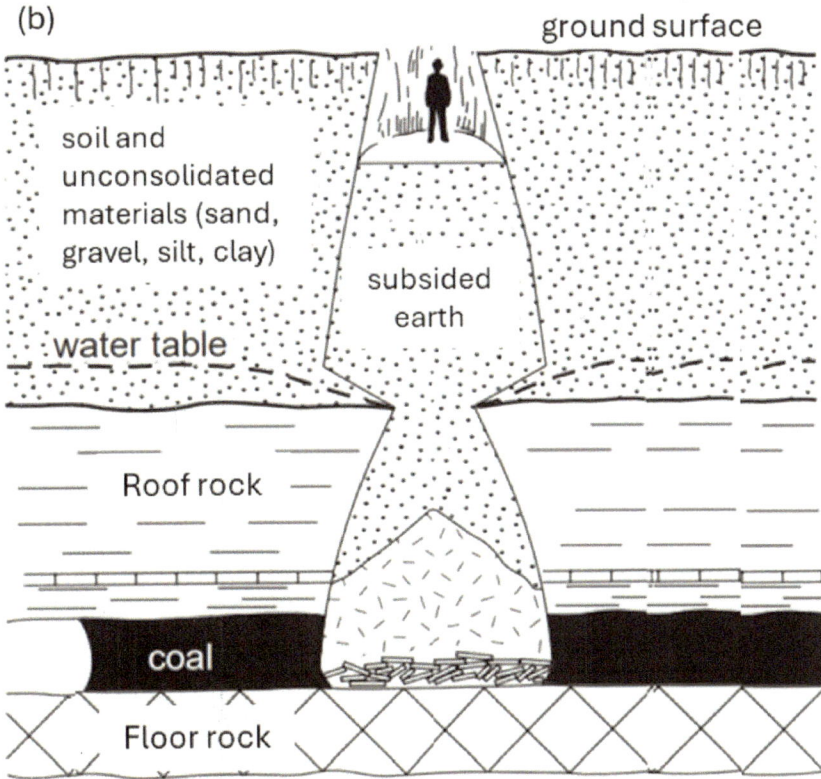

Fig. 1.1 Schematic examples of pit subsidence induced by mining (no scale implied)

Fig. 1.2 Typical pit subsidence caused by mine collapse

which eroded the fill, opening a subterranean channel to the Mississippi River. The catastrophic inflow exceeded the pumping capacity and flooded the quarry for over a year, while simultaneously triggering sinkhole collapses in the surrounding area, one of which caused a railroad wreck. The sinkhole formation processes related to dewatering in mines are schematically illustrated in Fig. 1.3. A combination of factors including soil weight, loss of buoyancy, suffusion process, and vacuum suction can contribute to the sinkhole formation.

Pit subsidence can also be associated with solution mining, which involves the process of dissolving soluble rock such as salt or potash. Figure 1.4 shows an example of such a mine pit. Solution mining typically creates large underground cavities that are filled with brine (Johnson 2005). The cavities typically are 10–100 m in diameter and are 10–600 m high, both dimensions being based largely on the thickness of the salt dome and the depth to the top of the cavity. Such cavities may end up quite long and relatively narrow when fresh water is pumped down one borehole and brine is extracted from the other.

It is possible for cavities to become larger or shallower than planned as a result of uncontrolled dissolution or unanticipated geologic conditions or engineering/ construction problems, and a number of these cavities have produced surface subsidence or collapse structures. Dunrud and Nevins (1981) reported ten areas of solution mining and collapse within the United States alone, and additional sites are known from many other parts of the world. Most unanticipated solution-mining collapses result from cavities formed 50–100 years ago, before modem-day engineering safeguards were developed.

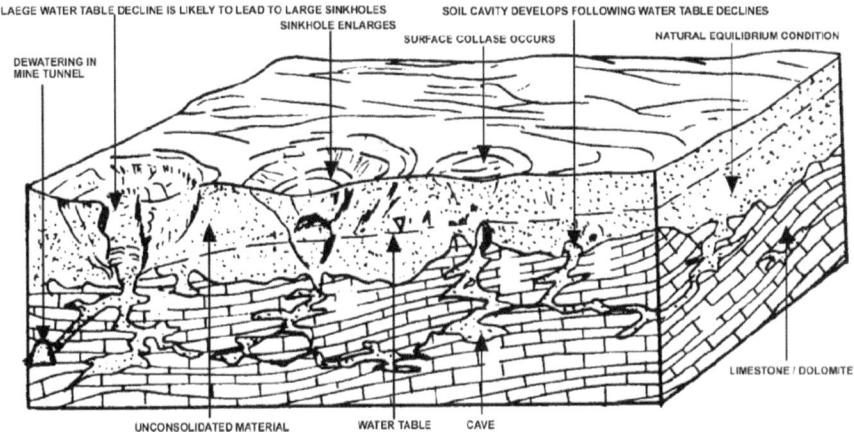

Fig. 1.3 Example of mine pits (sinkholes) associated with dewatering in underground mines

Fig. 1.4 Diagrammatic collapse (sinkholes) associated with solution mining (no scale implied)

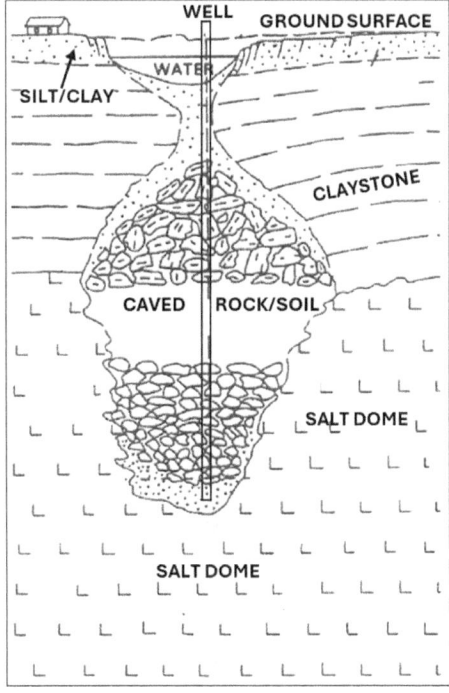

1.1.2 Trough Subsidence

Trough subsidence, also referred to as subsidence basin or sag subsidence, is a disc-
shaped depression that develops above the mined-out area and progressively enlarges
horizontally and vertically as coal support is systematically removed from beneath.
Such a subsidence typically occurs in a thick overburden. The resultant surface
effect is a large, shallow yet broad depression that is usually elliptical or circular
in shape. Long-wall mining, especially high-intensity long-wall mining tends to
create trough subsidence. The subsidence is usually greatest at the center of the
trough, and it progressively decreases until the limit of the impacted surface area
is reached. Horizontal ground movements also occur within a subsidence through.
Figure 1.5 shows a schematic diagram of the trough subsidence and deformation of
the overlying strata. An actual shallow depression caused by underground mining is
shown in Fig. 1.6.

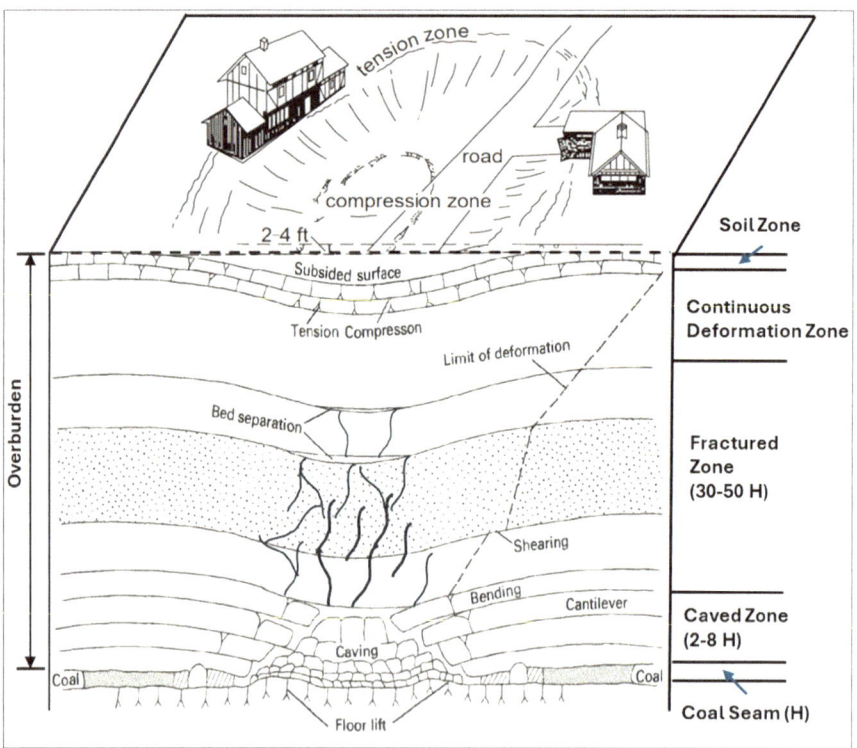

Fig. 1.5 Subsidence basin overlying a portion of a typical longwall panel showing four zones

Fig. 1.6 Typical trough subsidence caused by underground mining

According to the movement characteristics of the overlying strata, the impacted overburden can be divided into four zones:

- Caved Zone: This is the first zone above the unsupported void. After the extraction of coal, the immediate roof caves irregularly and fills up the void. The strata in this zone not only lose their continuity; they also lose their stratified bedding. The caved zone is normally 2–8 times the mining height depending on the properties of the immediate roof and the overburden.
- Fractured Zone: This zone is located immediately above the caved zone. The basis characteristics are strata breakage, and loss of continuity, but the stratified bedding remains. The severity of the strata breakage reduces from the bottom to the top. The porosity and permeability of the strata will increase greatly. The combined height of the fractured zone and caved zone is in general 20–30 times the mining height. The height of the fractured zone for hard and strong strata is larger than that for soft and weak ones. Peng (2006) has reported that the zone of extensive fracturing can extend from 34 to over 52 m above Pennsylvania longwall panels with coal mining thickness of 2 m.
- Continuous Deformation Zone: This zone is composed of the formation between the fractured zone and the surface bend downward without breaking. Their continuity and thus the original features remain. There may be open fissures in the tension zone of the surface subsidence profile that do not destroy the strata continuity. Bed separation voids can be created in layered strata within this zone.
- Soil Zone: This zone consists of soil and weathered rocks. Depending upon the physical properties of the soil, cracks may appear when the face is nearby and

close back when the face has passed. However, some cracks and especially those along the edges of the panel may remain open after mining.

1.2   Mine Subsidence as a Geohazard

Mine subsidence is a geohazard that can result in catastrophic and costly damage and even fatalities. Although some subsidence incidents are anticipated in mining areas, they can strike with little or no warning. Ground movement is typically swift and sudden for pit subsidence, while trough subsidence, which can develop over mines of any depth, appears as a gentle depression in the ground and can spread over an area as large as several square kilometers. The first signs may appear suddenly within a few hours or days, with gradual movement continuing anywhere from a few years to decades. Frequently, damage is less visible, and perhaps dismissed until multiple signs appear. Cumulative changes in surface slope, differential vertical displacements, and horizontal strains result in the damages from surface subsidence. The receptors to which mining-induced subsidence poses significant risks include structures, hydrogeological systems, plants, animals, and human beings.

Structures near the center of the trough can experience damage caused by the compression of the ground surface, and structures near the edges can be damaged by tension or stretching of the surface. Delayed subsidence can cause extensive damage in urban areas established over coal mines. The economic impacts of subsidence in rural areas can also be significant, and farm fields must be re-graded to eliminate ponding of water. Water wells may become dry when aquifers are disturbed by rock movements. Linear engineering structures such as roadways, railways, and gas mains are especially vulnerable to subsidence. Safety problems for residents caused by sinkholes and subsidence initiated by abandoned underground mines are a growing concern.

The trough-like subsidence areas formed over areas of longwall mining can cause damage to plant life and associated wildlife. Losses of soil water or water in deeper aquifers through fractures induced by subsidence could be harmful to plant and animal life, especially in arid and semiarid areas. Conversely, if these depressions are impermeable, they may fill with water, creating swampy, tree-killing conditions.

In addition, the flow of streams may be altered or disrupted, and surface cracks may occur, particularly near the edges of the trough. Formation of water ponds along a stream valley has been observed over a longwall mining area. The formation of stream ponds may cause problems for both the surface environment and the underground mining operations. The large body of water that formed on the surface may eventually enter the mine workings, posing a safety risk to the mine operations.

1.3 Extent of Subsidence Caused by Longwall Mining

The overlying strata sags downward and reaches the surface to form a low area of certain size above the gob. The concept of a "critical" area (width and length) of extraction is closely related to the ability of the strata above the excavated area to support loads across the mined openings. The formation of the trough subsidence begins when the gob exceeds a subcritical size. Of many factors such as mining method, depth of extraction, size and configuration of openings, rate of advance or extraction, seam thickness, topography, lithology, structure, hydrology, in situ stresses, and rock strength and deformational properties, the most important factors that affect the initiation and maximum lateral extent of subsidence are extraction width and mining depth.

1.3.1 Lateral Extent of Subsidence

A subsidence basin can be initiated when the ratio of the extraction zone width to the mining depth (overburden thickness) exceeds a value which varies from 0.1 to 0.5 with an average of 0.25 (Peng 1993). Factors such as the strength and structure of the rock overlying the mined-out area control the ratio value because of presence of a stabilizing compression (or pressure) arch in the solid rock above the mined-out area. The duration of this arching effect is controlled by the height, width, and length of the mined opening, and subsidence will not begin until the void size is so large that the arch spanning the excavated area is no longer stable. For this reason, a delay is often observed between the onset of a change in state underground and the first appearance of land subsidence. The arching effect may be compromised by very weak overburden rocks, groundwater flow conditions, or by poor mining practice that significantly weakens the overburden. Geologic conditions, mining depth, and seam thickness also affect arching behavior.

Figure 1.7 shows schematically the surface effects from longwall mining. On average, many longwall panels have an extraction zone width to overburden ratios ranging from 1.2 to 2.0, so well-formed subsidence basins are common. These ratios will yield supercritical subsidence basins. Subcritical widths of extraction produce a trough-like subsidence area with vertical subsidence less than the maximum. When the mining panel width exceeds a critical value, the subsidence reaches its maximum possible value and begins to spread, rather than increases further with increasing panel width. The ratio of critical width of extraction to mining depth ranges from 1.0 to 1.4 in European coal fields. This range has been attributed to differences in the types of overlying rock. The lower values of the depth coefficient appear to be associated with overburden containing thick, strong sandstone and limestone beds, whereas the higher values pertain to overburden containing a large percentage of thin-bedded shales, mudstones, siltstones, sandstones, and unconsolidated deposits.

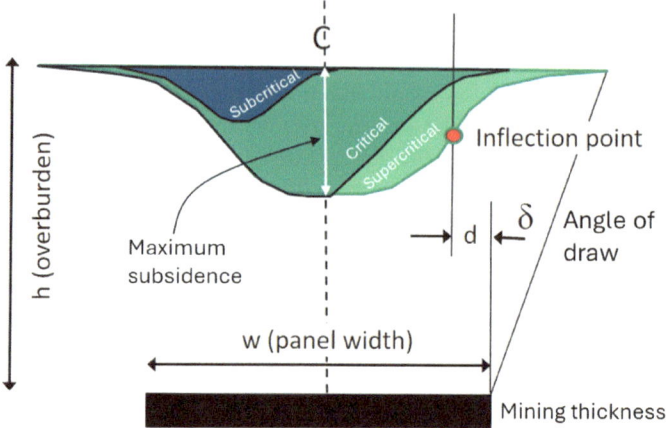

Fig. 1.7 Subsidence dynamics in response to longwall mining (d is the offset of inflection point)

Maximum subsidence is also a function of the thickness of the extracted layer, or the volume of material extracted, and the mining methods.

At supercritical widths of extraction, the subsidence area has an essentially flat bottom at approximately the maximum subsidence.

A trough is generally characterized by stationary surface profiles in the longitudinal and transverse directions and by non-stationary "dynamic" ground surface profile ("traveling wave") that accompanies the mine face in its passage from one end of the mine panel to the other. The maximum lateral extent can be approximated by mining depth (h) multiplied by the angle of draw (α), as expressed below:

$$L = 2(H_0 - h) \tan(\alpha) + 2h \cot \varphi \qquad (1.1)$$

where

H_0 mining depth of working face
α angle of draw in bedrock
h thickness of unconsolidated soil layer
φ angle of draw in soil layer.

The angle of draw (α) is defined as the angle formed between the edge of the longwall panel and the place on the surface where no measurable subsidence occurs. The prominent bending promotes separations along bedding as the strata moves inward toward the center of the subsidence basin. These fractures and bedding plane separations can affect the water-bearing strata by altering the groundwater flow path. The zone of groundwater impacts is most commonly found within the basin defined by the angle of draw. It outlines the limits of deformation associated with the subsidence basin formation. The affected ground will lie typically 10–45° outside the vertical limits of the mined area with the angle of draw being greater in bedrock than that in soil.

The developing subsidence basin produces a wave of deformation that first extends then compresses points on the surface. The transition from tension to compression occurs at the inflection point. The inflection point occurs in a continuous form around the developing dynamic subsidence basin. All points on the surface, between the side inflection points, experience both extension and compression. This flexing of the surface has the potential to impact all features overlying this portion of the longwall panel.

1.3.2 Vertical Extent of Subsidence

Factors that affect the lateral extent of mining-induced subsidence also affect the vertical extent. The most important factors that affect the vertical extent of the subsidence in coal mines are the mining thickness, extraction width, and mining depth. The relationship is well established in many coal mining areas between mining thickness and mining depth to vertical displacement of the ground surface. Mining of thinner seams or mining in shorter extraction width created smaller depth of the subsidence basin. Figure 1.8 show the relationship between percent subsidence and panel width for several extraction ratios in Illinois, United States (Hunt 1979). These examples would be typical trough-like subsidence. Where coal is extracted by room-and-pillar methods, attempts are usually made to control subsidence by leaving a larger percentage of the coal in place, often 40% or more. On the other hand, the modern longwall mining method tends to remove 100% of the coal along a straight working face within defined panels, up to several kilometers long and several hundred meters wide. The mine roof collapses immediately behind the moving roof supports, causing ground subsidence of 0.6–0.7 times the mined height of the coal seam over the centerline of the panel. It appears that the longwall mining has the greatest impact on subsidence.

These curves present an approximate range of control of mining method on subsidence. There is an obvious conflict between coal recovery and subsidence with larger resource recovery resulting in greater subsidence. A high percentage of coal needs to be left in the ground in order to obtain a high degree of surface stability. For example, the room-and-pillar method has been used with success when mining beneath towns, factories, railroads, and utility lines. The panels were mined without backfilling, and little subsidence damage occurred. Subsidence ratios were less than 20% of the coal thickness above mined panels where the widths were about one-third the overburden thickness. Therefore, if subsidence is not tolerable such as in urban areas or near important infrastructures, the coal recovery rate should be less than 50% of the resource. In many European coalfields where overburden thicknesses range from 60 to 900 m, deformation arches are stable within the overlying rocks, and subsidence does not reach the surface if the widths of mined-out areas are held from one-fourth to one-half of the overburden thickness (Lee and Abel 1983).

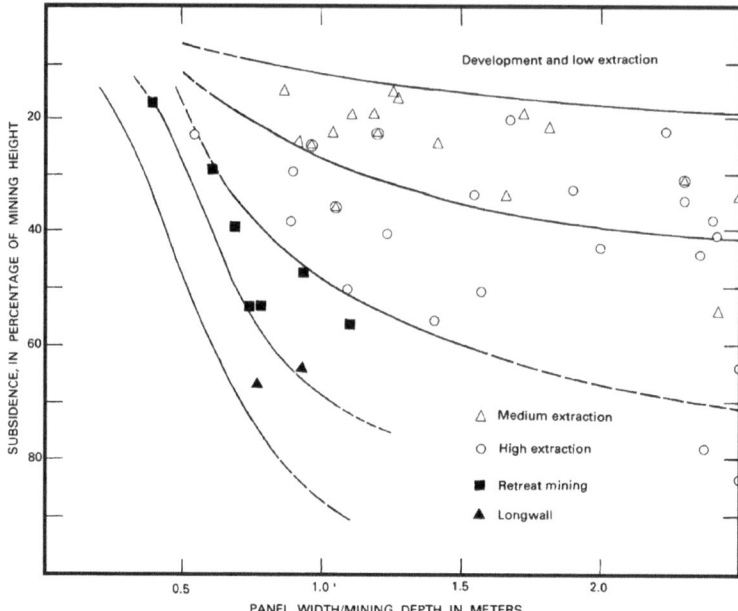

Fig. 1.8 General relation between percent subsidence and panel width to mining depth ratio (Hunt 1979)

Figure 1.9 shows the relationship between subsidence and depth to width based on the data collected by National Coal Board (1975) from more than 150 measurements made above longwall mines in the United Kingdom. For extraction width greater than 1.4 times the mining depth, the greatest possible subsidence is approximately 0.9 times mining thickness. The maximum subsidence for an extraction width of 0.7 times the mining depth is 0.45 times the mining thickness. The maximum subsidence for an extraction width of less than 0.2 times the mining depth is less than 10% of the seam thickness extracted. It appears that vertical extent estimated from Fig. 1.9 is similar to the estimated from Fig. 1.8 in coal mines where the extraction is greater than 80%.

The conditions under which Fig. 1.9 is applicable include the following:

- The overburden rock is largely shale, siltstone, or marlstone.
- The vertical component of surface displacement or subsidence (S) is associated with trough subsidence caused by the longwall mining extraction.
- The subsidence data are obtained from mining seams that dipped less than 25° and their thickness ranges from 0.6 to 5.5 m.
- The mining depth ranges from 30 to 792 m.
- The face or panel width ranges from 30 to 457 m.
- The panel width to depth ratio varies from 0.05 to 4.0.

Fig. 1.9 Empirical estimation of vertical extent of subsidence

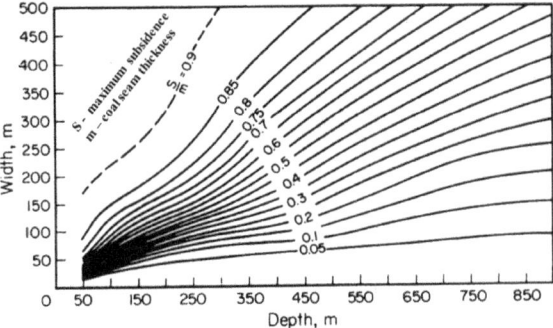

1.4 Subsidence Caused by High-Intensity Longwall Mining

Modern coal mines use the high-intensity longwall mining method to increase production efficiency. Such a mining method also has a great degree of disturbance and damage to the overlying strata of the coal seam, thus a great impact on the overlying aquifers and surface subsidence. Studies have been conducted with an attempt to strike a balance between extraction of resources and environmental protection including minimization of ground subsidence in mining areas. In-situ monitoring and statistical analysis methods have been used to understand the response characteristics of surface landform and stratum structure to the process of high-intensity mining (Li et al. 2018; Xu et al. 2020).

Zeng et al. (2023) did a focused study on impacts of high-intensity mining on overlying formation and ground surface in a working face of a coal mine in an arid and semi-arid area of west China. The width of the working face is 260 m. The advancing length is approximately 6004 m. The average mining depth is 318 m. The average thickness of the coal seam is 11.8 m, and the coal seam inclination angle is 0.4°.

1.4.1 Subsidence Measurements

The surface movement was monitored by establishing monitoring stations and measuring their elevations periodically in the course of mining and after. A total of 40 measuring stations were established along the advancing direction, while a total of 41 stations were established perpendicular to the advancing direction. The spacing between measuring stations was 15 m. Figure 1.10 shows the layout of the monitoring stations.

The elevations at these observation stations were measured periodically for a total of 18 times. The dynamic settlement curves over the working face are shown in Fig. 1.11. Figure 1.11a shows the subsidence processes along the advancing direction, whereas Fig. 1.11b shows the subsidence processes along the profile perpendicular

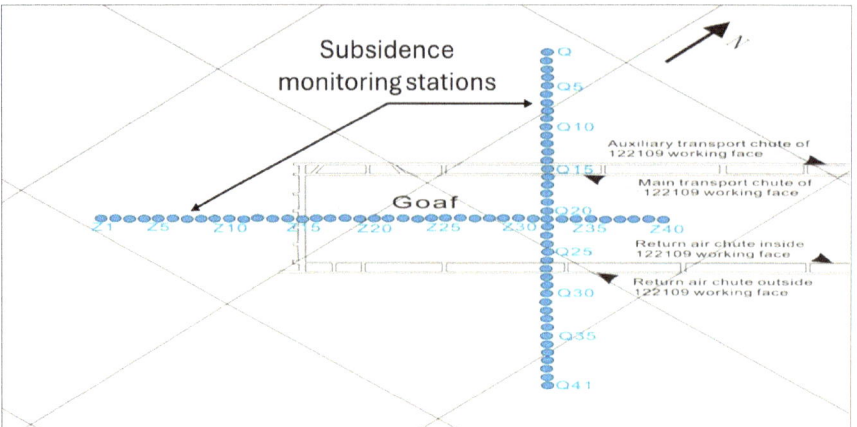

Fig. 1.10 Subsidence monitoring stations

to the advancing direction. As the working face advanced, the degree of the ground surface displacements continued to increase, and finally reached a stable state. The maximum subsidence value was 5.63 m, which is approximately 0.48 times the coal seam of 11.8 m.

The maximum subsidence occurred at the location where the survey lines crossed each other. The field measurements could not be performed at stations that were underwater because of ponding during the later stage of mining, resulting in some measuring points. A photo log was recorded for surface changes in each period when the elevation measurements were collected. The photos of the first and sixteenth periods are shown in Fig. 1.12 for the same location. The surface landform experienced disturbance and destruction with water accumulated in low-lying land. Vegetations were depressed and dead, and the land was no longer cultivatable.

1.4.2   Fractured Zone Measurements

The ground subsidence is directly related to the fractured zone induced by the high-intensity mining. The fractured zone height was monitored at completion of the mining over the same panel. The layout of monitoring boreholes is shown in Fig. 1.13. The distance of monitoring boreholes LD-1, LD-2, LD-3 is 2420 m from working face cut. The distance between LD-1 and the middle of the transport chute is 132 m. LD-2 is 20 m away from the transport chute. LD-3 is located in the middle of the coal pillar. LD-4 is 2270 m away from the working face cut and 20 m away from transport chute.

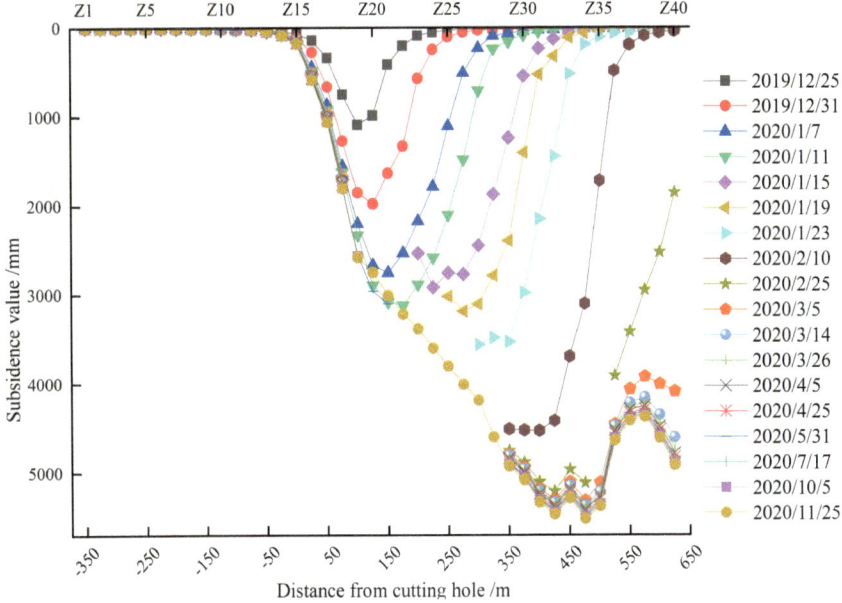

(a) **Profile along advancing direction**

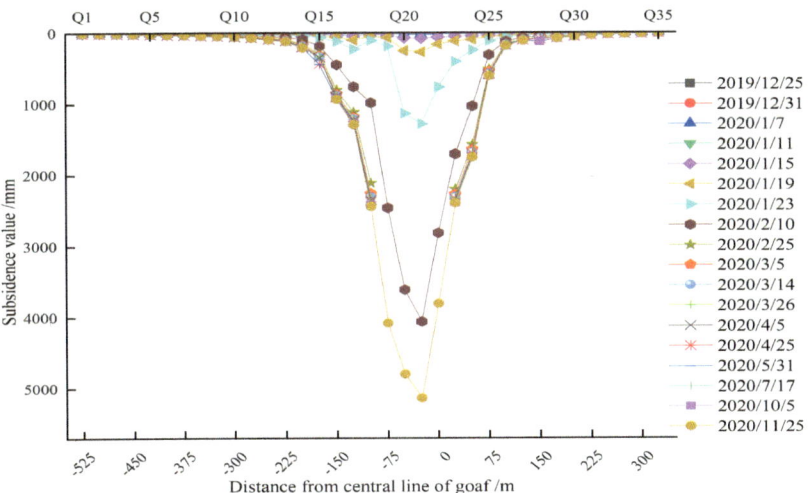

(b) **Profile perpendicular to advancing direction**

Fig. 1.11 Dynamic subsidence curve over the working face

(a) **Landscape in the 1ˢᵗ monitoring period**

(b) **Landscape in the 16ᵗʰ monitoring period**

Fig. 1.12 Landscapes deformation associated with underground mining

The water levels were measured in the four monitoring boreholes during the drilling process. Figure 1.14 shows the water level changes with the drilling depth. The water level in the borehole decreased with the drilling depth until it intercepted the top interface of the fractured zone where the water loss occurred. The hydraulic conductivity in the fractured zone is significantly enhanced by fractures induced by mining. The water level could no longer be detected. The detection results are summarized in Table 1.1. On average, the height of the water flowing fractured zone is approximately 21 times the mining thickness of the coal seam.

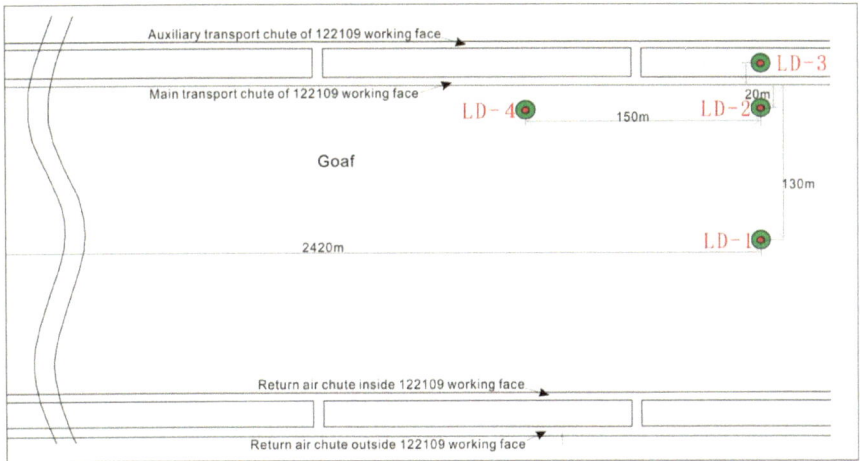

Fig. 1.13 Layout of monitoring boreholes

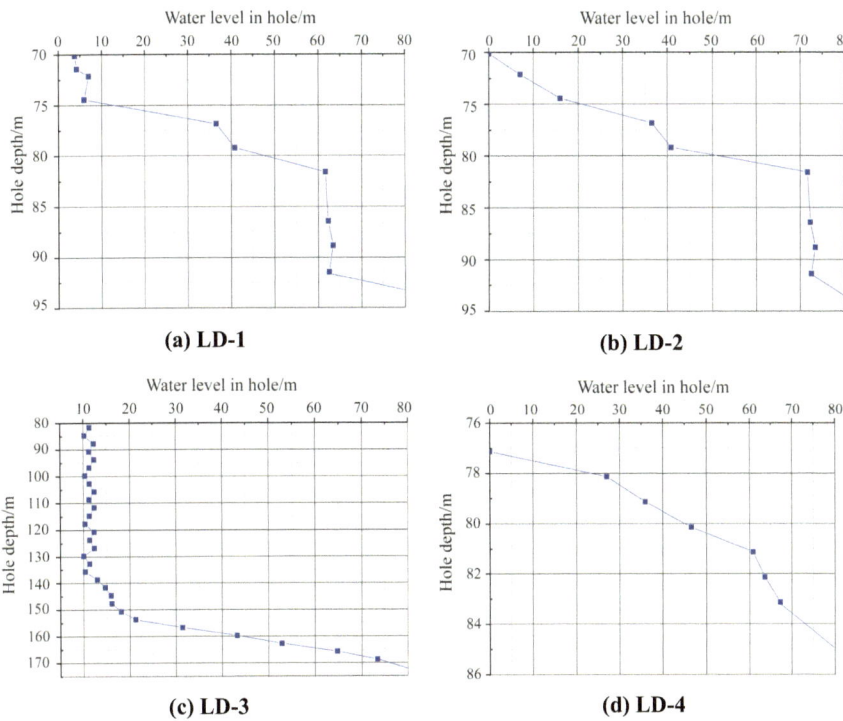

Fig. 1.14 Groundwater level changes during drilling

Table 1.1 Summary of detection results in monitoring boreholes

Monitoring borehole	Mining thickness (m)	Fractured zone height (m)	Ratio of fractured zone height to coal seam thickness
LD-1	11	210.05	19.09
LD-2	10.5	213.37	20.32
LD-3	11.5	140.52	12.21
LD-4	10.5	215.01	20.48

- The depth to complete water loss was 91.45 m below ground surface at borehole LD-1. Because the coal seam roof at LD-1 was 301.50 m below ground surface, the height of the fractured zone at LD-1 was approximately 210.05 m. The height also includes the caved zone.
- The water level at LD-2 borehole was still detectable at 72.76 m, and the circulation fluid completely lost at 86.7 m. The buried depth of the coal seam roof of LD-2 was 300.07 m, and the mining thickness of the coal seam at LD-2 working face was 10.5 m. As a result, the height of fractured zone at LD-2 was 213.37 m.
- The water level change at LD-3 borehole started from 74.95 m, and the circulation fluid completely lost at 168.6 m. The buried depth of the coal seam at LD-3 was 309.12 m, and the mining thickness of the coal seam at LD-3 was 11.5 m. The height of the fractured zone at LD-3 borehole was calculated to be 140.52 m.
- The water level observation at LD-4 borehole started from 77.76 m, and the water level was no longer detectable at 84.05 m. The buried depth of the coal seam at LD-4 was 299.06 m, and the mining thickness of the coal seam was 10.5 m. Therefore, the height of the fractured zone was 215.01 m.

These detection results indicate that the mining has caused damage to bedrock formations overlying the 2–2 coal seam. The top of the fractured zone was connected to the bottom of the laterite layer. There is a strong possibility that the fractured zone is hydraulically connected to the shallow porous media aquifer, especially at locations where laterite layer is thin and even missing. Under such a circumstance, not only does the ground subsidence become worse, but also can mine inrush accidents occur because water in the shallow porous medium aquifer can flow directly to the working faces. On the other hand, any engineering measures or optimization in mining design that help reduce the fractured zone height will certainly help alleviate the ground subsidence.

1.5 Subsidence Remedial Analysis

Subsidence remediation involves improving the landform to be useful and mitigating underground voids to prevent the subsidence from occurring in the future. The primary objective of a remedial analysis is to develop and evaluate options that will protect the public welfare, property, and the environment.

- Protect Public Welfare—by minimizing unacceptable hazards, preventing or reducing interruption of essential services, ensuring continued and uninterrupted access to emergency services, and preventing engulfment from catastrophic collapse. Alternatives that protect public welfare may include (1) administrative or institutional controls to restrict access to or use of the property such as zoning restrictions, installation of barrier fences and signage, and deed restrictions; (2) site monitoring systems to validate ground stability and, if needed, to provide early warning of impending collapse and allow time for evacuation; (3) engineering solutions that either mitigate sinkhole or enable nearby structures to withstand stresses from ground subsidence; and (4) provision of the data monitoring and early warning systems needed to ensure the safety and welfare of residents in the immediate vicinity of the sinkhole site.
- Protect Property—by characterizing and understanding the subsurface conditions to minimize or prevent a catastrophic collapse, preserving infrastructure, and ensuring that surrounding properties maintain current their uses. Damages to properties and infrastructures can be mitigated through engineering solutions, such as reinforcing and stabilizing the sinkhole to reduce rock stress and related ground subsidence.
- Protect the Environment—by preventing unacceptable risks to groundwater resources, mitigating the impact on existing habitat near the collapse area to the extent possible.

Selection of preferred remedial alternatives is based on three types of criteria: effectiveness, implementability, and relative cost.

- Effectiveness Criteria

  - Long-term treatment of hazards: Technologies with significantly lower long-term costs are preferred.
  - Ability to achieve the objectives to protect human health, property, and the environment: Technologies that would not effectively contribute to the protection of public health or the environment are not considered.
  - Permanent reduction in public hazard: Technologies that permanently reduce the hazard to the public are preferred.
  - Hazards to the public, workers, or the environment during technology implementation: Technologies posing less hazard during implementation are preferred.

- Implementability Criteria

  - Site characteristics limiting the construction or effective functioning of a technology: Technologies limited by site conditions are eliminated from consideration.
  - Subsurface characteristics that limit the use or effective functioning of a technology: Technologies limited by media characteristics are eliminated from consideration.
  - Availability of equipment needed to implement a technology: Commercially developed technologies that are readily available or innovative technologies that have been pilot-tested are preferred.
  - Administrative feasibility of obtaining permits and approvals from regulatory agencies and other offices: A technically feasible option may be difficult or impossible to obtain a permit for or to comply with the substantive requirements of the permit process. Technologies are eliminated if the permitting process is judged to be prohibitively difficult.

- Relative-Cost Criteria

  - Criteria for evaluating remedial technologies: Evaluations of technologies are qualitative, based on engineering judgment unless otherwise noted.
  - Process options within a technology: Relative magnitude of capital costs and operation and maintenance costs are compared.
  - Process options: If the effectiveness and implementability criteria are comparable, lower-cost options are preferred.

Table 1.2 presents an example on comparative analysis of select subsidence remedial alternatives. These alternatives provide only the basic principles of subsidence remediation and are not engineering designs. Table 1.3 summarizes the typical engineering measures. Ground subsidence in mining areas frequently presents "difficult ground conditions" to engineers, and its complexity is often inadequately appreciated. Depressions are a surface symptom of often complicated subsurface deformation processes induced by underground mining. They vary in size, shape, and type. Proper land use in subsidence-prone areas should include a subsidence management plan that prevents subsidence from occurring and avoids environment degradation. Various techniques are available for subsidence remediation, and their choice depends on the physical, hydrological and ecological properties of the subsidence and tolerance of the construction to collapse or subsidence. The engineering measures need to be tailored to address the specific conditions of the subsidence to be repaired. In general, subsidence remediation consists of plugging of the fractures, filling of the depression, and capping of surface. If the subsidence is improved to continue to drain surface runoff, it may be necessary that both the water quality and quantity need to be addressed. Institutional controls and monitoring are often two important components of long-term management for a subsidence site.

Table 1.2 Comparative analysis summary of select subsidence remedial alternatives

Alternative	Threshold criteria	Balancing criteria					Modifying criteria	
	Overall protection of public welfare, property, and the environment	Long-term effectiveness and permanence	Reduction of hazards, areas impacted, or severity of impacts	Short-term effectiveness	Implementability	Cost	Stakeholder acceptance	Community acceptance
(Symbols selected in the table are for demonstration purposes only. sinkhole-specific conditions shall be used to determine the type of symbol for each cell)	Parameters considered • Human health risk assessment • Public safety • Ecological risk assessment	Parameters considered • Residual risk at completion • Long-term management of remaining risks • Reliability of engineering controls and land use controls • Need to replace components • Continuing repair/ maintenance needs	Parameters considered • Hazard reduction • Impacted areas • Severity of impacts	Parameters considered • Short-term risks to community • Impacts on workers • Environmental impacts • Duration of remediation	Parameters considered • Technical feasibility • Operational reliability • Future alternative remedial options • Ability to monitor effectiveness • Ability to obtain governmental approvals • Availability of services and materials	Parameters considered • Net present value • Capital costs • Operation and maintenance costs	Parameters considered • Concerns of stakeholders • Regulatory compliance	Parameters considered • Concerns of community members • Ordnances

(continued)

Table 1.2 (continued)

Alternative	Threshold criteria	Balancing criteria						Modifying criteria	
	Overall protection of public welfare, property, and the environment	Long-term effectiveness and permanence	Reduction of hazards, areas impacted, or severity of impacts	Short-term effectiveness	Implementability	Cost	Stakeholder acceptance	Community acceptance
No action	○	○	○	○	●	●	○	○
Land use control with monitoring	◑	◑	◑	◑	◑	◑	◑	◑
Engineering control with monitoring and land use control	●	●	●	●	○	○	●	●

● Favorable; ◑ moderately favorable; ○ not favorable

Table 1.3 Commonly used engineering measures for subsidence remediation

Engineering measures	Description
Backfilling	• Most intuitive and simplest solution to remediation of an existing subsidence • Relatively a quick solution for constructions that can tolerate minor subsidence, and reactivation of the pits will not cause significant property damage and will not threaten people's life • Collapse throat exists and can be exposed by excavation. The excavation process should make sure that the pit throat is free of clayey materials. A grout plug is probably the most effective approach to remediate a subsidence pit • Any clay coating along the throat should be removed before concrete placement to secure a good bond between the concrete and rock
Fracture filling and backfilling	• Subsidence has no obvious throat but consists of many discrete fractures • These fractures can be impermealized by dental infilling grout • Pressure wash is recommended to help identify the fractures at exposed bedrock surface at the collapse bottom and to ensure that the grout pockets are free of clay • The pockets are filled with high/low slump flowable fill to plug and cap the fractures
Compaction grouting	• Typically applied to soil overburden or/and shallow rock within or near depressions • Relatively high grout pressure of 1380 kPa or greater is used in the grouting holes to fill voids, plug fissures around each hole, and displace/improve the soil/rock within a depression • Primary grouting hole spacing is typically 3.5–5 m. Smaller spacings may be needed for pinnacled rock • May cause additional fractures due to hydro-fracturing if it is not designed properly • May not work well in wet silts and clays • Can inadvertently seal off conduits that may be groundwater passageways
Jet grouting	• Involves pumping a fluid grout into the soil with a rotating high-pressure jet • The jet erodes soil and cuts stiff clays and soft erodible rock into gravel to small boulder-sized pieces • Pressures of 30–50 MPa are typical at the grout nozzle. The pressure dissipates rapidly within the soil and does not cause heave when the volume of the grout is properly controlled • The larger particles of soil, including sand and gravel in the sinkhole filling, mix with the grout, producing a mixed-in-place concrete

(continued)

Table 1.3 (continued)

Engineering measures	Description
Cap grouting	• Depressions associated with small but discrete fractures at the bedrock surface and the area to be treated is extensive • Cap grouting uses low grout pressures, 140 kPa or less to pump lean cement at the bedrock surface to impermealize the collapse bottom, fill voids, plug fissures, and displace soft soil • This operation provides support to the upper layer and disconnects any vertical hydrologic connections. Improvement of overburden soil is not obvious • Grout hole spacing is typically 0.9 m. In general, cap grouting does not consume as much grout as compaction grouting • Requires closer grout hole spacing for good coverage to intersect subsidence features • Auger drilling may not extend to bedrock due to shallow refusal on floaters • Hydro-fracturing possibility is limited; however, ground surface elevation should be observed for heave
Dynamic compaction	• Used to densify granular soils to depths of approximately 3–12 m • Involves dropping heavy weights, 5–30 short tons, on the soils from a pre-determined depth • Vibro-compaction uses larger diameter depth vibrators to densify and strengthen granular soils or to create stone columns in mixed or layered fine grained soils
Slurry grouting	• Involves injection of various mixtures of very fluid grouts into the ground • Fills cavities at virtually any depth that can be drilled. It can run along planes of weakness in the limestone and overburden forming very effective seals • Little to no densification of overburden soil takes place. Potential remains for building settlement
Filling with compacted clay	• To create an impermeable body to stop both vertical and lateral water flow • The fill consists of compacted clay with permeability lower than 10^{-6} cm/s • Recommend thin layers of soil because the compaction pressure transmitted to the soil decreases with soil depth • Compaction breaks down most aggregated or flocculated clay particles or vertical structure of the soil, making them less permeable

(continued)

Table 1.3 (continued)

Engineering measures	Description
Filling with graded filter	• Inverse aggregate graded filter • Place boulders wider than half the throat opening width into the fractures to arch across the bottom opening. Successive layers are sized finer than the underlying layer but coarse enough not to pass through the interstitial spaces of the bed beneath • Permeable from top to bottom when needed • Often permeable at the lower section and impermeable at the upper part to allow subsurface water moving at the soil/bedrock interface to access the drain and groundwater recharge to occur • Geotextile filter fabrics and even cement-grout mixes may be used when the stone layers are emplaced
Construction of water plugs	• Applicable to mines where depressions function as pathways for surface water to enter underground mines, causing serious damage and casualties • Use grout-plugs to eliminate hydraulic connections within the collapse feature • Grouting is conducted on surface by drilling boreholes and pumping grout into the collapse column, while monitoring of the effectiveness of the plug is conducted underground by artesian flow tests through directionally drilled wells • Completed grout-plugs range from tens of meters to hundreds of meters • A key component to the success of water plug construction is that all the grout holes including primary, secondary and tertiary ones are directed precisely at the designed locations • Effectiveness of completed plug is typically evaluated by core samples collected at boreholes and underground water-injection and water drainage tests

References

Abkemeier TJ, Stephenson RW (2003) Remediation of a sinkhole induced by quarrying. In: Beck BF (ed) Sinkholes and the engineering and environmental impacts of karst: proceedings of the ninth multidisciplinary conference, Huntsville, Alabama, 6–10 Sept 2003. American Society of Civil Engineers, Huntsville, pp 605–614

Dunrud CR, Nevins BB (1981) Solution mining and subsidence in evaporite rocks in the United States: U.S. geological survey miscellaneous investigations series map 1–1298, 2 sheets, scale 1:5,000,000

Foose RM (1967) Sinkhole formation by groundwater withdrawal: Far West Rand, South Africa. Science 157(3792):1045–1048

Hunt SR (1979) Characterization of subsidence profiles over room-and-pillar coal mines in Illinois. In: Illinois Mining Institute, 86th annual meeting, proceedings, pp 50–65

Johnson KS (2005) Salt dissolution and subsidence or collapse caused by human activities. In: Ehlen J, Haneberg WC, Larson RA (eds) Humans as geologic agents, vol 16. Geological Society of America Reviews in Engineering Geology, Boulder, Colorado, pp 101–110, https://doi.org/10.1130/2005.4016(09)

Lee FT, Abel Jr JF (1983) Subsidence from underground mining environmental analysis and planning considerations. Geological survey circular 876. U.S. Geological Survey, 28 pp

Li WP, Liu SL, Pei YB, He JH, Wang QQ (2018) Zoning for eco-geological environment before mining in Yushenfu mining area, northern Shaanxi, China. Environ Monit Assess 190(10):619. https://doi.org/10.1007/s10661-018-6996-5

National Coal Board (1975) Subsidence engineers' handbook. Mining Department of United Kingdom, London, 111 pp

Peng SS (1993) Surface subsidence engineering. SME, Littleton, CO, pp 52–61

Peng SS (2006) Longwall mining, 2nd edn. West Virginia University, Morgantown, West Virginia

Xu YY, Ma LQ, Yu H (2020) Water preservation and conservation above coal mines using an innovative approach: a case study. Environ Earth Sci Energies 13(11):2818. https://doi.org/10.3390/en13112818

Zeng YF, Pang ZZ, Wu Q, Lian HQ, Du X (2023) Roof water disaster in coal mining in ecologically fragile mining areas: formation mechanism and prevention and control measures. Springer, Cham. https://doi.org/10.1007/978-3-031-33140-4

Chapter 2
Investigations of Mining-Induced Subsidence

Abstract Mine subsidence and the factors that affect the subsidence vary significantly in different regions and different mines. Any investigations are to be designed to meet project objectives and tailored to site-specific conceptual site models. The requirements on data resolution typically increase as the project progresses from preliminary phase through planning phase to design and construction phases. During the preliminary and planning phases of a project, the primary objective is to evaluate the mine subsidence risk levels, whereas the primary objective of design and construction phases is to determine the depth, thickness, engineering properties of soil and rock, and hydrogeological conditions under the project site so that any potential of subsidence risk can be mitigated. As more linear projects such as roadways, railways, tunneling, and gas and oil pipelines cross large areas and new detection technologies are developed, selection of investigation techniques is not as distinctive between study scales. Many methods and techniques such as borehole exploration, groundwater monitoring, geophysics, tracer tests, and aquifer tests are applicable to in mine subsidence investigations. The level of effort depends on the complexity of the project site and severity of subsidence. It is a common practice that non-intrusive tasks precede the intrusive ones.

Keywords Borehole exploration · Geophysics · Tracer tests · Conceptual site model · Groundwater monitoring

2.1 Introduction

Mine subsidence and the factors that affect the subsidence vary significantly in different regions and different mines. Any investigations are to be designed to meet project objectives and tailored to site-specific conceptual site models. The requirements on data resolution typically increase as the project progresses from preliminary phase through planning phase to design and construction phases. During the preliminary and planning phases of a project, the primary objective is to evaluate the mine subsidence risk levels, whereas the primary objective of design and construction

© The Author(s), under exclusive license to Springer Nature Switzerland AG 2024   27
S. Dong et al., *Prevention and Reclamation of Mining-Induced Land Subsidence*,
SpringerBriefs in Earth Sciences, https://doi.org/10.1007/978-3-031-78158-2_2

phases is to determine the depth, thickness, engineering properties of soil and rock, and hydrogeological conditions under the project site so that any potential of subsidence risk can be mitigated. As more linear projects such as roadways, railways, tunneling, and gas and oil pipelines cross large areas and new detection technologies are developed, selection of investigation techniques is not as distinctive between study scales. Table 2.1 presents the main components that are often included in mine subsidence investigations. The level of effort of each component depends on the complexity of the project site and severity of subsidence. It is a common practice that non-intrusive tasks such as literature review, subsidence reconnaissance, and geophysical surveys precede the intrusive ones such as drilling and aquifer testing.

The dynamic conditions of groundwater should be monitored to collect the essential data during the investigation phase. The ground deformation monitoring and subsurface soil erosion monitoring should be arranged in subsidence-prone areas or areas where the engineering activity is intense.

2.2   Preliminary Investigation

Preliminary investigation includes a desktop study and an inventory of mine subsidence features. The purpose of the preliminary investigation is to identify areas of concern that may require additional investigation and to review the preliminary site design in relationship to potential problem areas. The preliminary site investigation may result in immediate changes to the site layout to avoid future problems. Desktop study relies on collection and analysis of existing data. The recommended mapping scale is 1:50,000. The inventory of subsidence features consists of not only collection of existing data but also field reconnaissance.

2.2.1   Desktop Study

Desktop study consists of collecting and reviewing historical documents from various agencies, consulting forms, and published records of the study area. The fundamental data include the following three profiles:

Physical Profile

- Meteorological data
- Topographic maps including aerial photograms and remote sensing maps
- Vegetative types and patterns
- Surface water features and drainage pathways
- Engineering geological conditions, including soil type and geotechnical properties
- Hydrogeological conditions such as depth to groundwater and characteristics of aquifers and subsidence features recognized

Table 2.1 Components of mine subsidence investigation

Investigation components	Objective	Preliminary phase	Planning phase	Design and construction phases
Mine geology and hydrogeology mapping	Collect existing data, understand mining history, and plot hydrogeologically significant data on maps	Target map resolution		
		1:50,000	1:500–1:10,000	1:200–1:500
Subsidence reconnaissance	Field reconnaissance on subsidence features and receptors affected by subsidence	✓	✓	✓
Geophysical survey	Non-intrusive investigation for site characterization along transects or borehole loggings		✓	✓
Drilling exploration	Intrusive investigation for site characterization, target hydrogeologically significant features or geophysical anomalies		✓	✓
Trenching	Intrusive investigation for overburden features		✓	✓
In-situ test	Soil testing on strength parameter or permeability in field		✓	✓
Geotechnical test of rock and soil sample	Laboratory testing of soil and rock samples to determine geotechnical properties		✓	✓

(continued)

Table 2.1 (continued)

Investigation components	Objective	Preliminary phase	Planning phase	Design and construction phases
Tracer test	Field investigation for site characterization to understand groundwater flow and hydraulic connections		√	√
Pumping or slug test	Field investigation for site characterization to understand hydraulic properties of the underlying aquifer		√	√
Monitoring of hydrodynamic conditions	Field investigation to collect data on groundwater or gas pressure in mining influenced formations and soil formations, and precipitation		√	√
Monitoring of ground surface deformation	Field investigation to collect data on ground movements in both horizontal and vertical directions		√	√
Monitoring of fracturing and deformation in subsurface soil and bedrock	Field investigation to collect data on soil erosion and bedrock fracturing process		√	√

- Naturally occurring phenomena (e.g., tidal action or earthquake) that may cause subsurface fracturing.

Human Activity Profile

- Mining history including quantity of resource extraction, mining methods, and backfilling activities
- Underground utility distribution including water mains, sewer systems, and storm drains
- Past, current, and anticipated future land uses

- Ground disturbance activities such as surface drainage modification, investigation, and construction
- Dewatering in mines or pumping in wellfields.

Subsidence Profile

- Subsidence size
- Subsidence shape
- Geologic formation in which Subsidence occurred
- Type of subsidence (pits or troughs)
- Stage of subsidence
- Imminent threat to safety and the environment.

Caution should be exercised when using standard topographic maps to identify subsidence-prone areas because the number of depressions depicted on the maps are related to the contour intervals. It is typical that more depressions are identified on the ground than were depicted on the maps. However, not all closed topographic depressions are depressions. Depressional features that are not mine subsidence include old quarrying pits, borrow pits, test pits, and dug drainage ponds, which can all have circular to semi-circular map profiles and have closed topographic depression elevations.

Use of satellite and remote sensing data has become an important part of preliminary investigations. High-resolution satellite images and aerial remote sensing maps such as digital elevation model (DEM), light detection and ranging (LiDAR), InSAR provide useful information on surface landform, morphological types, and mine subsidence distribution. When aerial maps are available at various times, changes in conditions of the investigation area can be evaluated. Closed depressions identified on aerial maps are typically researched using geographic information system (GIS) and serve as a planning tool that helps guide systematic and efficient verification in field reconnaissance of mine-related subsidence.

2.2.2 Subsidence Reconnaissance

Subsidence reconnaissance is an essential step in any mine subsidence study. Figure 2.1 shows an example subsidence reconnaissance form. The purpose of the reconnaissance is to compile and plot all reconnaissance features. Surface features such as pits, sinkholes, and troughs are cataloged along with pertinent information such as size, shape, location and orientation for subsidence and location. If concentric earth fissures are found, these are also noted. Other features of geologic interest such as outcrops, and any important information that they may provide such as the strike and dip of the bedrock, or the location and attitude of faults and fractures, are also recorded. Well locations are collected if logs are available, or water levels can be recorded. Data on these features are then tabulated and their locations plotted on a map or aerial photograph for further analysis. A synthesis of this data and its distribution can be used to formulate a preliminary conceptual model of the subsidence development. The information can also be used to guide additional research efforts such as geophysical investigations and groundwater tracing.

The term subsidence is applied to a variety of related features, which can be confusing. From a geological point of view, internally drained trough depressions are not the same type of features as the steep-sided pits that form by collapse. It is important to understand why and how these two types of subsidence develop. Water flowing down a pit may feed directly to the underground mine. The location and distribution of pits provide important information about the mining conditions

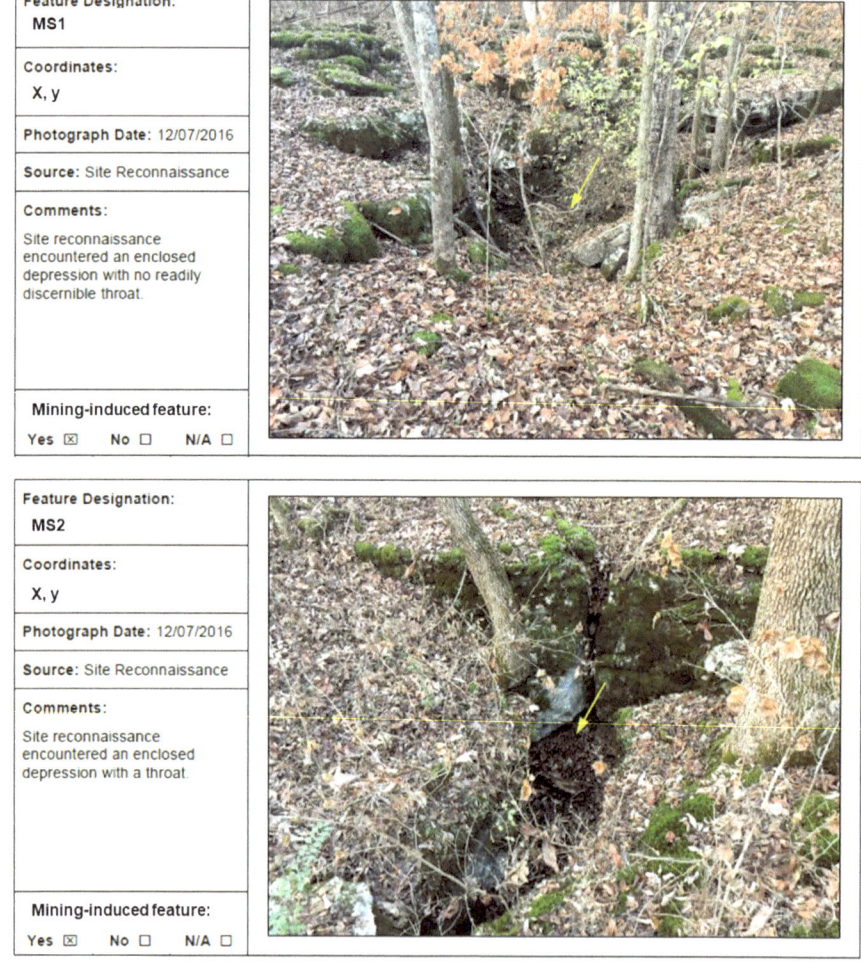

Fig. 2.1 An example subsidence reconnaissance form

and groundwater flow conditions. In a subsidence reconnaissance, in addition to documenting depressions that have complete topographic closure, i.e., once water crosses the topographic threshold it cannot flow out, the following signs of surficial deformation should be noted:

- Vertical to sub-vertical soil cracks concentric to depression's perceived center point which may create a complete to partial ring around the depression
- Sagged or sagging ground in relation to the near-vicinity ground surface topography
- Soil creep or slumped or slumping soil
- Arcing trunks of trees and shrubs attempting to re-straighten to vertical orientation within the depression perpendicular to depression's perceived center point, indicating soil creep
- Trees, shrubs, or other vertical features that are leaning or sagging into the depression
- Exposed rock or semi-indurated sediments which otherwise would not be exposed at near-vicinity ground surface topography
- Water marks on foliage indicating the depression is actively internally draining
- Water flow marks on ground which orient sediments or foliage litter towards the lowest elevation(s) within the depression
- Vegetation showing signs of stress or dying within the depression.

The subsidence reconnaissance should differentiate mining-induced depressions from the natural ones. Natural subsidence refers to depressions formed under natural conditions, and the triggering factors include rainfalls, earthquakes, or long-term dissolution of bedrock. Induced subsidence refers to the depressions caused by mining or dewatering in mines and tunnels, reservoir leakage, blasting, drilling, or loading.

Interview of local people or those who have institutional knowledge of the site is part of subsidence reconnaissance. The following are typical questions:

- Knowledge of any water ponding on the property and indicate any changes in the water volume.
- Knowledge of any stream located on the property and indicate if the stream flow appears to diminish with distance.
- Knowledge of any depressions located on the property.

Data collection involves assigning each subsidence feature a unique identification number and surveying its coordinates. Coordinates should be approximated if the subsidence feature is based anecdotal information. Subsidence features are catalogued and presented in tabular form with subsidence feature designation, coordinate information, photographic documentation, and comments. Features identified on historical documents that field reconnaissance is not able to confirm due to lack of a surface expression should be recorded as previously noted. Features are depicted on aerial plan sheets with a unique symbol assigned to each type of subsidence feature.

Table 2.2 Classification of mine subsidence scales

Classification index	Classification			
	Extra-large scale	Large scale	Medium scale	Small scale
Subsidence diameter (m)	> 60	20–60	10–20	< 10
Number of depressions	> 20	10–20	5–10	< 5
Affected area (10,000 m²)	> 10	5–10	1–5	< 1

The scale classification should be errored to larger scales, and the number of depressions include collapse pits and troughs

Based on size of depressions, number of depressions, and area affected by subsidence, mine subsidence can be divided into four scales: extra-large, large, medium, and small (Table 2.2).

2.2.3 Results of Preliminary Investigation

Data collected during the preliminary investigation should help develop the preliminary conceptual site model of mine subsidence and evaluate vulnerability of mine subsidence. The preliminary conceptual site model should address the following components:

- Intensity, development rate, and frequency of the mine subsidence
- Formation conditions and main control factors of mine subsidence
- Inducing factors, their dynamic changes, and their relationship to mine subsidence
- Composition and distribution characteristics of hydrogeological features and dynamic variation characteristics of groundwater level and permeability
- Relationship between morphological characteristics and distribution patterns of mine subsidence
- Development phases of mine subsidence should be ensured
- Preliminary classification of mine hydrogeological condition according to Table 2.3.

2.3 Site Characterization During Planning, Design, and Construction Phases

Site-specific investigation, including investigation components during planning and design or construction phases, is conducted once the decision is made to design a site plan and proceed with development. The main scope of work is similar for these phases, but the data resolutions are different. Requirements on survey line spacing and data density along each survey line are more stringent during design and construction phases. Table 2.4 presents general guidelines on the level of effort.

Table 2.3 Complexity of mine hydrogeological conditions

Indicator	Complexity of geological conditions		
	Complex	Moderate	Simple
Relationship between fractured zone height and thickness of impermeable bedrock overlying coal seam	• Fractured zone height greater than thickness of impermeable bedrock, resulting in occurrence of pits and direct recharge from overlying aquifer or surface water	• Fractured zone height marginally less than thickness of impermeable bedrock. Mine pits or troughs can occur if the working face dimension is not designed properly	• Fractured zone height significantly smaller than thickness of impermeable bedrock
Intensity of mine subsidence	• Extensive depressions • Presence of depression groups • Large and extra-large trough subsidence	• Some depressions • Mainly single subsidence pits • Medium-sized subsidence trough	• Sporadic depressions • Small subsidence pits

Note Division of complexity is based on the principle of erroring to more complex

The target mapping scale for planning investigation is between 1:500 and 1:10,000, whereas the target mapping scale for design and construction investigation is between 1:200 and 1:500. The area of interest in the design and construction phases should encompass the risk-elevated areas that required prevention and control as delineated in the planning phase.

2.3.1 Geophysical Surveys

Exploration geophysics is the science of seeing into the subsurface without intrusive investigations. Geophysical methods can help to establish the geological setting of a mining environment and be used to identify hydraulically relevant structures, such as mined-out voids, caved zones, or fractured zones. A variety of geophysical methods such as gravity, electrical resistivity, or acoustic wave velocities are available based on different physical properties of the soil and rock. The major advantages of geophysics include relatively rapid and inexpensive site coverage compared to detailed drilling campaigns. While drillings deliver precise data for selected points, geophysics can generate less expensive but less precise data for larger areas. In mining areas, geophysics is particularly useful for identifying suitable locations for drilling of wells and for mapping overburden thickness, verifying burial and filling conditions of cavities, but also for geotechnical investigations, such as the identification of potential subsidence hazards below structures. Table 2.5 summarizes the applicable geophysical methods based on the objective of the survey. These geophysical

Table 2.4 Level of effort for mine subsidence investigation during planning, design and construction phases

Investigation components	Parameter	Planning phase	Design and construction phases
Field reconnaissance	Transect spacing (m)	50–500	30
	Data point spacing along transect (m)	20–100	15
Drilling exploration	Transect spacing (m)	30–500	15
Geophysical survey	Transect spacing (m)	30–1000	15
	Data point spacing along transect (m)	5–10	5–10
Monitoring of hydrodynamic conditions	Distance between monitoring points (m)	50–500	40
Monitoring of ground surface deformation	Distance between monitoring points (m)	20–500	10
In-situ test	For natural soil, the common methods of in-situ tests are static cone penetration test, dynamic penetration tests, standard penetration tests, and vane shear tests. In-situ tests are generally conducted in combination with drilling. Conduct at least 6 tests per soil type		
Geotechnical test and analysis of rock and soil sample	The common geotechnical indexes conclude water content, density, and specific weight of soil particles, grain composition, water ratio limit, shear strength, wetting-induced disintegration characteristic and expansion rate. Chemical analyst should be done for carbonate rocks. The content of CaO, MgO, SiO_2, and R_2O_3 and other constituents should be measured. Specific solubility and specific corrodibility could be measured if necessary. Collect at least 6 samples per soil type		
Geochemical analysis of water sample	Analysis of water samples is essential to determine the water chemical properties and groundwater type. Additional analytes should be added if necessary. Collect water samples at wells and springs as required by projects		
Tracer test	Tracer test is a viable tool to determine groundwater flow, velocity, and connections between depressions and underground working panels		
Pumping or slug test	Slug or pumping tests are commonly used to characterize aquifers and acquire hydrology parameters in select wells as required by project. Water injection tests could be arranged if site conditions are not in favor of pumping tests		

techniques and discussions of their applicability to mine subsidence investigation are based on the conceptual site model and physical properties of various geological media. Figure 2.2 shows an example of conceptual site model in which seven factors are associated with mine subsidence.

- Mined-out void left from coal extraction
- Caved zone immediately above the mined-out area

Table 2.5 Applicability of main geophysical methods to mine subsidence investigation

Investigation purpose	Viable geophysical method	Application condition
Overburden structure and thickness	Ground penetrating radar (GPR), seismic refraction, electrical resistivity imaging, seismic reflection	(1) Selection based on detection depth (D in meter) When D < 10 m: ground penetrating radar method, seismic refraction method, Rayleigh wave method
Depth to bedrock and bedrock undulation	Electrical resistivity imaging, Rayleigh wave seismic, seismic refraction, seismic reflection, ground penetrating radar, transient electromagnetic, audio frequency or controlled source audio frequency magnetotelluric	When 10 < D < 30 m: seismic refraction method, Rayleigh wave method, electrical resistivity imaging, seismic reflection method When 30 < D < 50 m: electrical resistivity imaging, seismic reflection method When D > 50 m: electrical resistivity imaging, transient electromagnetic method, audio or controllable source audio magnetotelluric method (2) Selection for anomaly positioning Multiple parallel transect resistivity imaging, induced polarization, natural potential, high precision micrography, radioactive method, microdynamic method (3) Selection of qualitative methods Water abundance: polarization method, microdynamic method (4) High-resolution positioning Borehole geophysical method: borehole radar, cross-hole electromagnetic method, cross-hole acoustic wave method, cross-hole resistivity imaging (5) Noise selection of geophysical methods based on noise interference Urban areas are characterized with dense building and strong electromagnetic and vibration interference. Recommended geophysical methods are GPR, high-precision micrography, micro-dynamics, cross-hole resistivity, and borehole radar In villages and towns, it is recommended to use GPR, electrical resistivity imaging, shallow seismic, micro-dynamics, CSAMT also known as V8, audio magnetotelluric AMT also known as EH4, cross-hole imaging, and borehole radar Geophysical prospecting in the borehole should be conducted whenever an opportunity arises, including cross-hole electromagnetic imaging, borehole sonar three-dimensional scanning, and borehole geological radar

(continued)

Table 2.5 (continued)

Investigation purpose	Viable geophysical method	Application condition
Soil cavity, soft soil, collapse	Ground penetrating radar, seismic reflection, microgravity, Rayleigh surface wave, electrical resistivity imaging, radioactive method	
Mine voids and fractured zone	Borehole geophysical logging, electrical resistivity imaging	
Strong water-transmissive zones	Induced polarization, natural potential, borehole geophysical logging, electrical resistivity imaging, transient electromagnetic, audio frequency or controlled source audio frequency magnetotelluric, air-borne geophysics	
Groundwater flow conditions	Induced polarization, natural potential, infrared thermography, air-borne geophysics	
Paleochannel	Electrical resistivity imaging, seismic reflection, transient electromagnetic, audio frequency or controlled source audio frequency magnetotelluric	
Weathered zone	Seismic refraction, electrical resistivity imaging, Rayleigh wave	
Interface between continuous deformation zone and fractured zone	Electrical resistivity imaging, transient electromagnetic, audio frequency or controllable source audio frequency magnetotelluric, induced polarization, microdynamic	

(continued)

Table 2.5 (continued)

Investigation purpose	Viable geophysical method	Application condition
Fault, fractured zone, caved zone and groundwater preferential flow zone	Electrical resistivity imaging, seismic refraction, transient electromagnetic, controlled source audio frequency magnetotelluric method, radioactive geophysical, microdynamic, induced polarization, borehole hydrophysical logging	
Rock mass loose zone, rock integrity, dynamic elastic mechanics parameters	Seismic refraction, acoustic logging	
Morphology and filling characteristics of mined-out areas	Ultrasonic logging, sonar logging, borehole acoustic or optical televiewer	

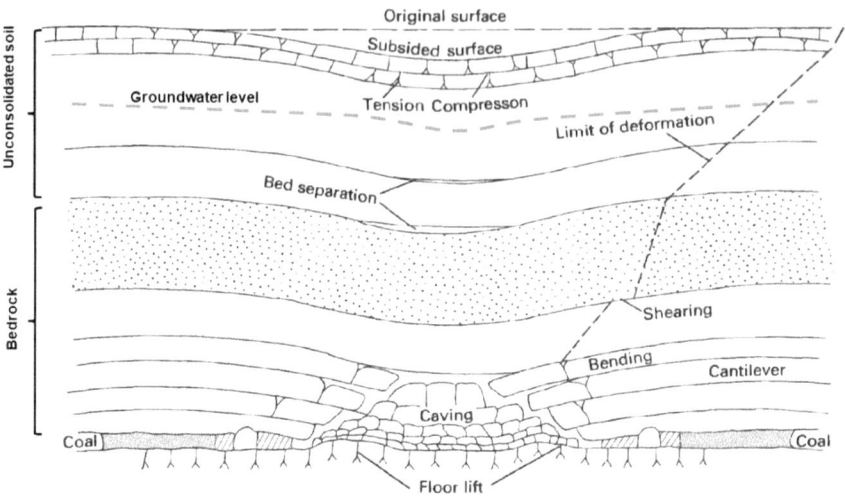

Fig. 2.2 Subsidence conceptual site model for geophysical application

- Fractured zone
- Voids from bed separation
- Continuous deformation strata
- Ground subsidence
- Potentiometric pressure.

The seven factors can be considered as variables and their specific features can change. For example, a new conceptual site model is developed if the overlying strata are not layered. Or, changing of the potentiometric pressure to different position results in another conceptual site model. Such changes affect the applicability of geophysical techniques. Table 2.6 presents the geophysical properties that are associated with different geophysical methods.

Electrical resistivity imaging or electrical resistivity tomography may be the most frequently used for site investigation in mining areas, especially when the overburden soil is clay-dominated. The electrical conductivity of clayey soil and carbonate rock has an electrolytic origin. Carbonate rock in general has a significantly higher resistivity than clayey soil because it has much smaller primary porosity and fewer interconnected pore spaces. Clayey materials tend to hold more moisture and have a higher concentration of ion to conduct electricity. The high contrast in resistivity values between carbonate rock and clayey soil favors the use of resistivity method to delineate the boundary between bedrock and overburden. Commonly used electrode geometry includes Wenner, Schlumberger, pole-dipole, and dipole-dipole. Of the commonly used arrays, dipole-dipole provides the highest resolution and is most sensitive to vertical resistivity boundaries (Zhou et al. 2002) as are found at undulating bedrock interfaces.

Table 2.6 Geophysical properties of geological media

Geologic medius	Resistivity (Ω m)	Relative dielectric constant	P-wave velocity (m/s)	Magnetic susceptibility ($4\pi \times 10^{-6}$)	Density (g/cm^3)
Clay	$10^1 \sim 10^3$	5 ~ 40	300 ~ 2000	0	1.6 ~ 2.6
Soft soil, silty clay	$10^0 \sim 10^2$	15 ~ 60	100 ~ 600	–	1.4 ~ 3.0
Silt	$10^1 \sim 10^3$	5 ~ 10	200 ~ 800	0	1.7 ~ 2.5
Wet sand, pebble	$10^2 \sim 10^3$	25 ~ 30	700 ~ 1600	0	1.7 ~ 2.5
Dry sand, pebble	$10^3 \sim 10^5$	2 ~ 6	300 ~ 1000	0	1.7 ~ 2.4
Clayey gravel	$10^2 \sim 10^3$	–	800 ~ 2500	–	–
Mudstone	$10^1 \sim 10^2$	–	3000 ~ 4000	–	–
Shale	$2 \times 10^1 \sim 10^3$	7	2300 ~ 4700	10	2.0 ~ 2.5
Argillaceous sandstone	$10^1 \sim 10^2$	9 ~ 11	1400 ~ 4300	10	1.6 ~ 2.7
Quartz sandstone	$10^2 \sim 10^3$	6	2400 ~ 4300	350	2.6 ~ 2.7
Coal	$10^{-1} \sim 10^2$	–	2500 ~ 3500	–	1.1 ~ 1.9
Argillaceous limestone	$50 \times 10^1 \sim 8 \times 10^2$	1 ~ 50	3500 ~ 4500	–	2.2 ~ 2.4
Limestone	$3 \times 10^2 \sim 10^4$	7 ~ 8	100 ~ 6400	5	2.2 ~ 2.9
Dolomite	$10^2 \sim 10^4$	7 ~ 8	2500 ~ 6200	0	2.8 ~ 3.0
Carbonaceous rock	$10^0 \sim 10^2$	8	3000 ~ 4000	–	–
Clastic tuff	$2 \times 10^2 \sim 10^3$	5	4000 ~ 5000	–	–
Granite	$2 \times 10^3 \sim 10^5$	5 ~ 7	4500 ~ 6500	500	2.5 ~ 3.0
Diorite	$5 \times 10^2 \sim 10^5$	8	4500 ~ 6500	1000	2.7 ~ 3.0
Basalt	$5 \times 10^2 \sim 10^5$	18.8	4500 ~ 8000	1000	2.7 ~ 3.0
Schist	$2 \times 10^2 \sim 10^4$	–	2500 ~ 5000	1000	–
Slate	$10^1 \sim 3 \times 10^2$	–	4000 ~ 4500	500	2.5 ~ 2.6
Gneiss	$2 \times 10^2 \sim 5 \times 10^4$	–	3000 ~ 5500	25	2.5 ~ 3.3
Marble	$10^2 \sim 10^4$	8.5	4500 ~ 5500	5	2.6 ~ 2.9
Halite	$10^4 \sim 10^8$	6	2400 ~ 6500	0	2.07 ~ 2.17
Debris	12 ~ 30	–	200 ~ 600	–	–

Figure 2.3 presents an electrical resistivity profile interpreted from resistivity imaging over an old mining area (Sheets 2014). Three distinct areas of very high apparent resistivity (> 1000 Ω m) are shown at approximately 26, 38, and 72 m (horizontal distance). These anomalous areas are about 3 m deep. The very high apparent resistivity is probably not related to natural geologic materials and is instead interpreted to be air-filled voids such as abandoned coal mine goafs or tunnels. The somewhat continuous sub-horizontal high-resistivity anomaly that spans the very high-resistivity anomalies from 38 to 72 m (horizontal) likely is a series of mine voids (tunnels) separated by sections of unmined coal. Geological loggings of the borings completed along the survey line are presented to show that the high-resistivity areas correspond to void locations found in the borings.

Data collected from earth resistivity imaging are easily affected by near-surface resistivity variations and therefore can produce noisy data at sites with cultural relics. In addition, three-dimensional variation in geology can be an important factor affecting the reliability of this technique. A comparison of the depth to limestone determined from pre-existing borings with that interpreted from resistivity imaging transects showed an average difference of 2.4 m, with a maximum of 10 m (Zhou et al. 2002). Averaging the interpreted elevations from several transects reduces the discrepancy.

Controlled source audio magnetotelluric (CSAMT) is a type of controlled-source electromagnetics surveying that involves transmitting a controlled electric signal at a suite of frequencies into the ground from a far field location (i.e., transmitter site) and measuring the received electric and magnetic fields in the study area (i.e., receiver site). It evaluates the earth's subsurface electrical resistivity distribution by measuring time dependent variations of the earth's natural electric and magnetic fields, as well as the electric and magnetic fields resulting from high frequency induced waves. Changes in the electric and magnetic fields over a range of frequencies enable an apparent resistivity sounding curve to be constructed. Apparent resistivity is combined with a measure of the phase difference between the electric and magnetic components. Joint inversion of the data using both phase and apparent resistivity provides a more robust interpretation. Modeled data is displayed as apparent resistivity versus depth plot. Factors affecting resistivity values include geologic structures, porosity, groundwater, and the presence of contaminants. Like earth resistivity imaging, CSAMT technique can provide critical information about geologic structure, lithology, water-table trends, and groundwater flow paths and is sensitive to low resistivity geological structures, such as fault zones. CSAMT is a more suitable geophysical technique at greater depth than earth resistivity imaging.

Woods and Marinello (1999) reported a case study in which a large solution mining void was delineated in a salt dome using a high-resolution magnetotelluric remote sensing system. Natural electromagnetic energy comprises the source for the system. The source is believed to be many orders of magnitude greater than the strengths of fields that can be generated with man-made sources on the surface of the earth. The energy originates in particle form as "solar wind" and is transduced into wave energy as it passes through the ionosphere and magnetosphere. Very low frequency (0.0005–100 Hz) carrier waves subsequently enter the earth, and the

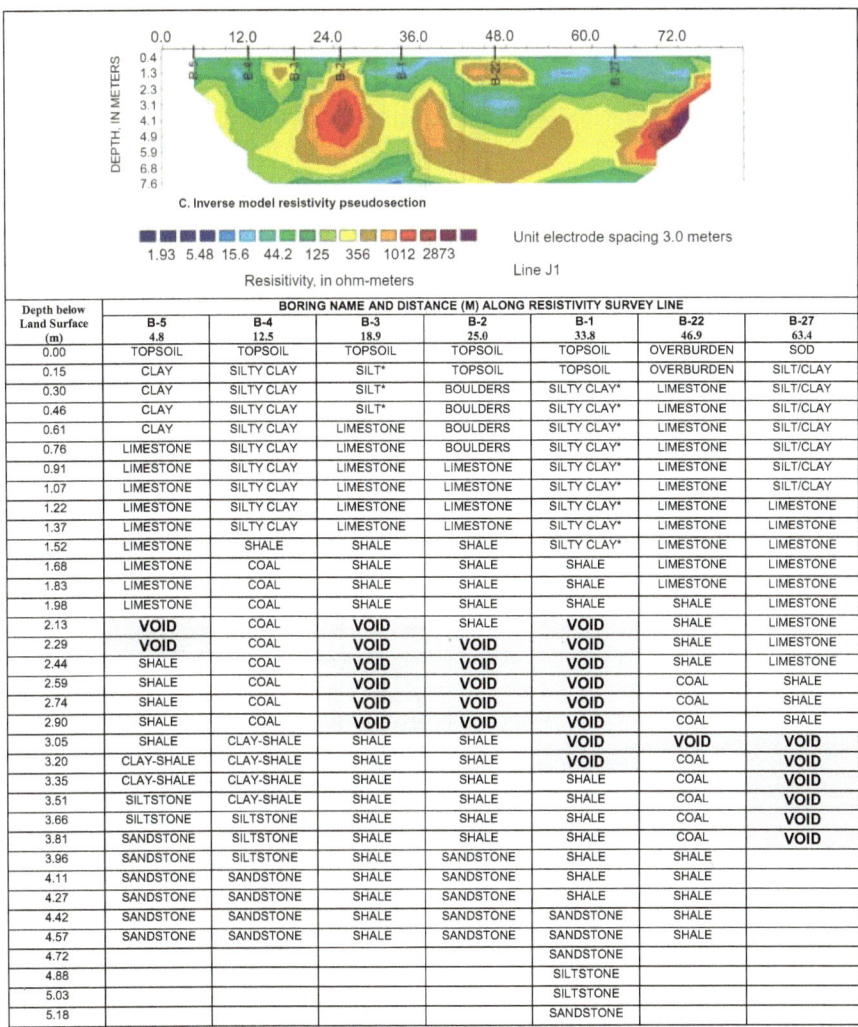

C. Inverse model resistivity pseudosection

Unit electrode spacing 3.0 meters

1.93 5.48 15.6 44.2 125 356 1012 2873

Resisitivity, in ohm-meters

Line J1

Depth below Land Surface (m)	BORING NAME AND DISTANCE (M) ALONG RESISTIVITY SURVEY LINE						
	B-5 4.8	B-4 12.5	B-3 18.9	B-2 25.0	B-1 33.8	B-22 46.9	B-27 63.4
0.00	TOPSOIL	TOPSOIL	TOPSOIL	TOPSOIL	TOPSOIL	OVERBURDEN	SOD
0.15	CLAY	SILTY CLAY	SILT*	TOPSOIL	TOPSOIL	OVERBURDEN	SILT/CLAY
0.30	CLAY	SILTY CLAY	SILT*	BOULDERS	SILTY CLAY*	LIMESTONE	SILT/CLAY
0.46	CLAY	SILTY CLAY	SILT*	BOULDERS	SILTY CLAY*	LIMESTONE	SILT/CLAY
0.61	CLAY	SILTY CLAY	LIMESTONE	BOULDERS	SILTY CLAY*	LIMESTONE	SILT/CLAY
0.76	LIMESTONE	SILTY CLAY	LIMESTONE	BOULDERS	SILTY CLAY*	LIMESTONE	SILT/CLAY
0.91	LIMESTONE	SILTY CLAY	LIMESTONE	LIMESTONE	SILTY CLAY*	LIMESTONE	SILT/CLAY
1.07	LIMESTONE	SILTY CLAY	LIMESTONE	LIMESTONE	SILTY CLAY*	LIMESTONE	SILT/CLAY
1.22	LIMESTONE	SILTY CLAY	LIMESTONE	LIMESTONE	SILTY CLAY*	LIMESTONE	LIMESTONE
1.37	LIMESTONE	SILTY CLAY	LIMESTONE	LIMESTONE	SILTY CLAY*	LIMESTONE	LIMESTONE
1.52	LIMESTONE	SHALE	SHALE	SHALE	SILTY CLAY*	LIMESTONE	LIMESTONE
1.68	LIMESTONE	COAL	SHALE	SHALE	SHALE	LIMESTONE	LIMESTONE
1.83	LIMESTONE	COAL	SHALE	SHALE	SHALE	LIMESTONE	LIMESTONE
1.98	LIMESTONE	COAL	SHALE	SHALE	SHALE	SHALE	LIMESTONE
2.13	VOID	COAL	VOID	SHALE	VOID	SHALE	LIMESTONE
2.29	VOID	COAL	VOID	VOID	VOID	SHALE	LIMESTONE
2.44	SHALE	COAL	VOID	VOID	VOID	SHALE	LIMESTONE
2.59	SHALE	COAL	VOID	VOID	VOID	COAL	SHALE
2.74	SHALE	COAL	VOID	VOID	VOID	COAL	SHALE
2.90	SHALE	COAL	VOID	VOID	VOID	COAL	SHALE
3.05	SHALE	CLAY-SHALE	SHALE	SHALE	VOID	VOID	VOID
3.20	CLAY-SHALE	CLAY-SHALE	SHALE	SHALE	VOID	COAL	VOID
3.35	CLAY-SHALE	CLAY-SHALE	SHALE	SHALE	SHALE	COAL	VOID
3.51	SILTSTONE	CLAY-SHALE	SHALE	SHALE	SHALE	COAL	VOID
3.66	SILTSTONE	SILTSTONE	SHALE	SHALE	SHALE	COAL	VOID
3.81	SANDSTONE	SILTSTONE	SHALE	SHALE	SHALE	COAL	VOID
3.96	SANDSTONE	SILTSTONE	SHALE	SANDSTONE	SHALE	SHALE	
4.11	SANDSTONE	SANDSTONE	SHALE	SANDSTONE	SHALE	SHALE	
4.27	SANDSTONE	SANDSTONE	SHALE	SANDSTONE	SHALE	SHALE	
4.42	SANDSTONE	SANDSTONE	SHALE	SANDSTONE	SANDSTONE	SHALE	
4.57	SANDSTONE	SANDSTONE	SHALE	SANDSTONE	SANDSTONE	SHALE	
4.72					SANDSTONE		
4.88					SILTSTONE		
5.03					SILTSTONE		
5.18					SANDSTONE		

Fig. 2.3 Example of resistivity survey and corroborated borehole data

amplitude and phase of individual frequencies are modified at resistivity interfaces from which they are reflected back to the surface. This system varies from classical magnetotelluric methods by analyzing a secondary set of higher frequency harmonics (100–22,000 Hz) generated by the large impedance contrast at the air-ground interface. These frequencies carry the same relative amplitude data as the returning carrier frequencies, but facilitate a significant improvement in resolution, allowing for detailed vertical subsurface sampling with depth errors of < 5 m, in most areas, with proper calibration. The technology employs the simultaneous recording of the electric (E) and magnetic (H) components of the telluric field. Through processing, the arrays of amplitudes of their frequency series can be measured separately, or as ratios (E/H). These values are displayed in vertical profiles of the subsurface for inter-pretation. Figure 2.4 shows an example of such a profile with brief interpretations. The void locations correspond to the low values of E/H ratio. Interpreting as strong water zones in this case. Zones annotated as weak water are thought to be either fluid filled rubble or fractured zones. The shape and areal extent of the underground void can be delineated by the strong water zones from a series of such vertical profiles.

Natural potential (NP) method is one of few geophysical techniques in which the anomalies are associated with groundwater flow rather than the mere presence of groundwater (Lange and Barner 1995). The measurement of NP can be used to characterize groundwater flow in underground voids because electrical potential gradients are generated by the horizontal flow of water along fractures or conduits and the vertical infiltration of water into fractures or shafts (Zhou et al. 1999). Identi-fication of vertical recharge zones is essential in evaluating risk of potential collapses for construction, waste disposal and other activities in mining areas. The most signif-icant NP results could be obtained where considerable quantities of water flow into collapse pits (Erchul and Slifer 1987).

Detection of the lateral and vertical extents of goafs is essential in subsidence investigations and remediation in mining areas. The applicability of the geophysical techniques varies and to a large extent depends on whether the goafs are flooded with groundwater or not. Table 2.7 summarizes the main physical properties of the goafs, whereas Table 2.8 summarizes selection criteria on geophysical techniques.

Non-uniqueness is a common problem of geophysical data interpretations. Different geological features or their combinations can cause the same geophys-ical anomaly, which means in turn that any observed anomaly can be interpreted in different ways. Non-uniqueness can be partly overcome by combining different geophysical methods and/or by checking geophysical interpretations by boreholes. Another general problem of geophysics is the trade-off between depth of investigation and resolution. The greater the depth, the lower the resolution.

Borehole geophysics or hydrophysics or borehole logging is an indispensable supplement to the analysis of subsurface conditions. Borehole geophysics makes it possible to conduct high-resolution characterization in situ. Temperature logs are a simple type of borehole geophysics and an example of a relevant underground property that cannot be obtained from drilling cores. Bechtel et al. (2007) summarize available borehole geophysical methods and their uses in hydrogeology studies.

One useful borehole logging tool in detecting water-filled mine voids is the sonar survey. Sonar, an acronym for sound navigation and ranging, is a technology that maps and locates objects by acoustic means. An echo-transducer (sender-receiver) generates a pulse of sound that reflects off the surface of an object, then records the return time of the pulse. The travel time of the sound wave to and from the object is proportional to the distance from the echo-transducer to the object. Under favorable conditions, this technology provides high-quality geometry data for open voids. A camera is usually also included to give visual data of the borehole (i.e., during descent into a void). Figure 2.5 shows a Sonar setup and detected image.

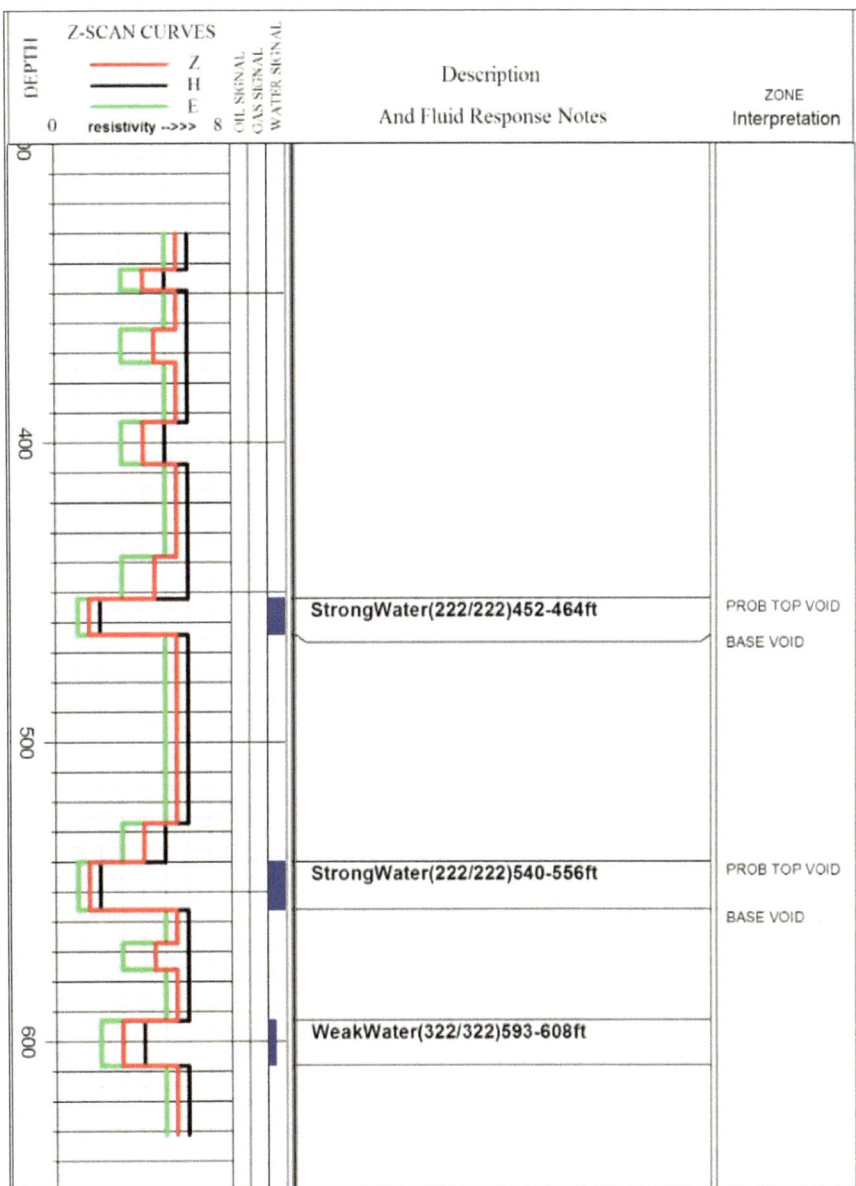

Fig. 2.4 Example profile from the high-resolution magnetotelluric remote sensing system

Table 2.7 Geophysical properties of goafs

Goaf type	Anomalous electrical resistivity	Anomalous elasticity	Anomalous radiation	Anomalous gravity	Anomalous cross-hole CT
Goaf with no water	Medium-high resistance, weak dielectric, low induced polarization, low electromagnetic attenuation, and high resistance relative to unmined coal seams	Low speed and low density, with a strong reflection arc-shaped wave group at the top or an interruption of the reflection wave common phase axis in the unmined coal seam, showing chaotic and obviously low-frequency reflection wave abnormal characteristics	The upper surface is rich in radon, and along its border there is a significant high radon anomaly	It has lower density and shows low relative gravity anomaly	Relatively high resistance, low density and low speed, showing excessive distribution of high resistance or low speed anomalies on the upper boundary
Goaf with water pools	Medium-low resistance, high dielectric, high induced polarization, high NMR amplitude, high electromagnetic attenuation, and low-resistance or low-frequency weak reflection chaotic wave anomalies relative to unmined coal seams	Relatively low speed and low density, the top presents a strong reflection arc-shaped wave group or the reflection wave common phase axis of the unmined coal seam is interrupted, showing weak, chaotic, low-frequency and low-amplitude reflection wave abnormal characteristics	The upper near-surface is radon-rich, with relatively high radon anomalies distributed along its boundaries	Medium-low density, weak low gravity anomaly	Relatively low resistance, low density and low speed, showing an excessive distribution of low resistance or low speed anomalies on the upper boundary

The various geophysical tools can contribute to an initial understanding of the general underground void geometry below the subsidence area or prior to subsidence occurrence. They also provide essential information in the development of a drilling program and limited knowledge of void properties and conditions. The drilling program provides continuous core samples for geological logging, rock samples for testing, and borehole access through which borehole geophysical logging can be performed. The borehole geophysical logging results are directly correlated with the geologic core logs and testing of samples. This 'ground truthing' of geophysical measurements in the nearby borings provides context for interpreting the geophysics. The geophysical methods are often not capable of reliably identifying specific voids. Direct evaluation of open void volumes is not feasible until the borings are advanced

Table 2.8 Selection of geophysical techniques for goaf detection

Goaf type	Defects of using one single method				Recommended combination of methods		
	Electrical resistivity	Seismic	Microgravity	Cross-hole CT	Shallow	Intermedia	Deep
Goaf with no water	• There is an electrical equivalence effect • The boundaries of the three-band anomalies are not obvious	• There are obvious scattering and absorption effects • Affected by terrain • High cost and low efficiency • It is not convenient to analyze the properties of the filling materials in the goaf	• Low anomaly intensity requires a certain scale of goaf • High accuracy is required for terrain correction • Quantitative interpretation is prone to multiple solutions	• Resistivity CT has an electrical equivalence effect • Elastic wave CT is not convenient for analyzing the properties of the filling material in the goaf	Ground penetrating radar or high-resolution resistivity and Rn measurement	Seismic reflection wave method and Rn measurement method	Seismic reflection wave or high-precision gravity exploration and Rn measurement
Goaf with water pools	• There is an electrical equivalence effect • There is a low resistance effect • The accuracy of the three-band boundary division is low	• There are obvious scattering and absorption effects • It is affected by the terrain • It is costly and inefficient • It is not convenient to analyze the properties of the filling materials in the goaf	• The anomaly intensity is low, which requires the goaf to be of a certain size • The terrain correction requires high accuracy • Quantitative interpretation is prone to multiple solutions	• Resistivity CT has an electrical equivalence effect • Elastic wave CT is not convenient for analyzing the properties of the filling material in the goaf	High-density resistivity method and Rn measurement method	Transient electromagnetic or seismic reflection wave method and Rn measurement method	Seismic reflection wave or frequency magnetotelluric method and Rn measurement

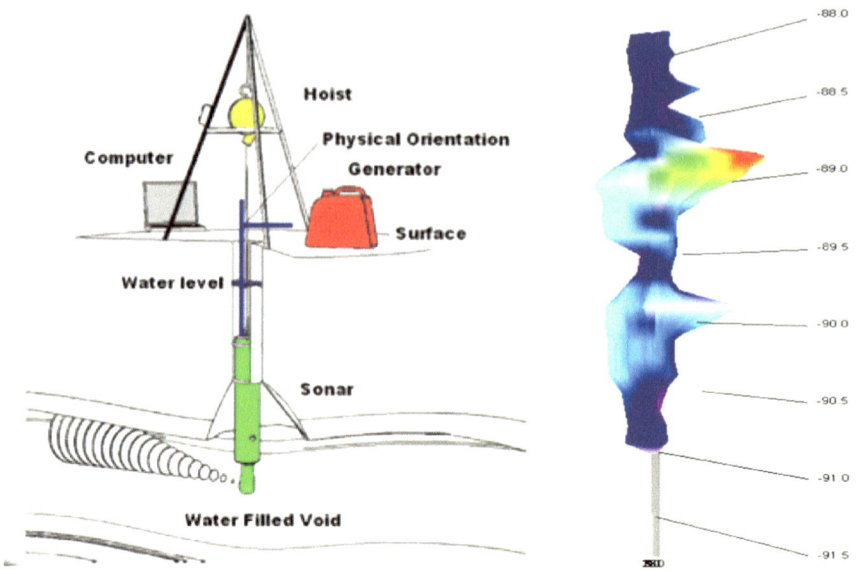

Fig. 2.5 Sonar survey setup and an image example of mine voids

into the cavity. Borehole-based methods that penetrate the cavity will better identify specific void locations and extent.

2.3.2 Drilling Exploration

Drilling is an essential investigation tool to obtain direct data of the subsurface. Objectives of drilling vary with project requirements, and some are provided below:

- Determine the lithology, thickness, structure, spatial distribution and variation of each rock and soil.
- Verify the lateral and vertical extent of mined-out areas or height of the fractured zone.
- Understand the shape, scale, filling situation and spatial variation of underlying fracture development.
- Determine hydrogeological property and hydraulic relationship between underground working area and overlying porous medium or fracture aquifer.
- Collect rock and soil samples and field tests to understand properties of rock and soil samples and their variations with depth.
- Install monitoring wells for hydrodynamic condition monitoring.
- Facilitate grouting operations to remediate subsidence areas or subsidence-prone areas.

The main drilling techniques include direct push technology (DPT), hollow stem auger (HSA), sonic, rotary, and directional drilling. DPT and HSA are applicable to soils at shallow depths, while sonic and rotary drilling are more applicable to bedrock at greater depths. To improve the accuracy of concealed subsidence detection, the use of DPT and HSA is encouraged to perform a larger amount of drilling work to determine the thickness, lithology, soil structure, and depth to auger refusal. Drilling integrated with standard penetration tests and cone penetrometer soundings provide specific and accurate information on soil and rock stratigraphy as well as soil strength and consistency, and location of voids and loose raveling zones within the overburden.

Because of heterogeneity in overburden and underlying formation, the number of borings required to characterize a site can vary significantly. Although Table 2.4 gives a recommendation on data point density, site-specific design is required. Mining changes hydrogeological and geotechnical properties of the overlying formations, and the as-built mining plan or records of mining history are critical in helping design the drilling program.

Drilling design for subsidence investigation and mitigation should recognize that fact that landscape in a mining area is different from the original landscape. Mining causes deformations in the overlying strata, resulting in caved zone, fractured zone, continuous deformation zone, and soil zone. A major challenge in mine subsidence investigations is finding vertical and lateral extents of the deformed formation. Although drilling design is site-specific, the following principles help place the drilling locations:

- Drilling is based on field reconnaissance and geophysical survey, and efforts are made to serve each borehole for multiple purposes such as determination of subsidence development characteristics and geotechnical characteristics of the study site, verification of geophysical anomalies, and installation of a monitoring station to provide real-time monitoring of ground settlement.
- Layout of drilling locations is designed to control different types of rock and soil, as well as being perpendicular to topography and geomorphology, structural lines, and direction of groundwater flow.
- For main subsidence areas with high density of collapses, boreholes should be arranged to control the formation conditions and expose the main fractured zone. The layout of the exploration line should be arranged along the expansion direction of the subsidence.

The drilling depth and final borehole diameter should meet the following requirements:

- Hydrogeological drilling is generally required to expose the main aquifer (group) or fractured zone. The depth of the borehole should generally expose the fractured development zone or main aquifer (group).
- For engineering geological drilling, in thin-covered subsidence areas, the borehole depth should reach 20 m below the bedrock surface; in thick-covered subsidence areas, the borehole depth could exceed 100 m.

- To meet sampling and testing requirements, borehole diameter in the soil layer is generally not less than 110 mm, whereas borehole diameter in the rock is not less than 91 mm. For boreholes for special tests, the borehole size should be determined according to the needs.

If core collection is required, the drilling should meet the following requirements:

- Continuous drilling is required. Soil layers are drilled using direct push technology or hollow stem auger, whereas carbonate rock and water-resistant layers are drilled with clean water.
- Core collection rates vary with soil types and rock formations: no less than 80% for cohesive soil and intact rock; no less than 70% for sandy soil, no less than 60% for soft soil, gravel soil, cave filling and broken zone. Core intervals shall not exceed 1 m except for cohesive soil, which requires core interval not exceeding 0.5 m.
- Samples are collected in more than 1/3 of total number of boreholes and taken in layers. The number of samples for soil layer, bottom soil layer, fractured zone, and caved zone filling should not be fewer than 6 groups.

During drilling, attention should be paid to observe the starting and ending depths of drill bit drops, decompression of drilling tool, and loss of circulation, and determine the starting and ending burial depths and filling levels of the intercepted fractured zone or caved zone (full filling, partial filling, no filling). The initial, stabilized, or loss of groundwater level of the borehole should be recorded. After the groundwater level is observed in the borehole, it is required to measure the water level in the borehole once every shift. The starting and ending depths should be recorded in which loss of circulation fluid occurs, water gushes out, and watercolor changes. When conditions permit, water temperature in the borehole should also be measured. When conducting a standard penetration test, the estimates of soil engineering characteristics including blow counts or estimated unconfined compressive strength should be recorded.

2.3.3 Tracer Test

Tracer testing can be a viable technology to investigate causes of ground subsidence in mining. There are several types of materials that can be used as tracers. Table 2.9 summarizes some commonly used tracers.

Use of tracers to determine hydrologic connection, direction of groundwater movement, and/or the flow rate can be challenging in mining areas because of the following:

- Use of the wrong tracer for the conditions and type of test to be performed.
- When insufficient tracer is introduced at the input point, the concentration may become so dilute by the time it reaches the measurement point it will not be noticeable or even measurable. This is especially true if a tracer is placed in an

Table 2.9 Commonly used tracers

Type of tracer	Commonly used tracers
Fluorescent (organic or optical) dyes	• Eosine—color index name "Acid Red 87" • Fluorescein—color index name "Acid Yellow 73" • Rhodamine WT—color index name "Acid Red 388" • Sulforhodamine B—color index name "Acid Red 52" • Pyranine—color index name "D&C Green 8"
Chemical tracers	• Sodium chloride—common salt, both the sodium and the chloride are relatively chemically conservative so both can act as tracer ions, if background concentrations are low • Calcium chloride • Bromide • Iodide • Fluoride
Isotope tracers	• Naturally occurring stable ($\delta^{18}O/\delta D$) • Radioactive (^3H) water isotopes
Gas tracers	• Helium • Neon • Argon • Xenon • Krypton

underground mine pool that may have tens of millions to more than a billion gallons of water in storage and/or flushing of the pool is usually relatively slow.

• Receiving a non-detect at a sample point does not necessarily mean that there is no connection because:

- The sampling intervals are too large to catch the passing of the tracer.
- The flow rate is so slow that the tracer has not arrived within the monitoring period.
- The tracer is diluted and attenuated below detection limits by the time it reaches the sampling point.

• The use of fluorescent dye tracers in mining situations has some limitations. The geochemical environment common to underground and surface coal mines has a tendency to adversely affect many optical dyes. Most fluorescent dyes will adsorb to iron hydroxides, clays, and organic materials. These materials are extremely common in coal mines and can be contacted by the mine water. Additionally, these dyes are negatively impacted by changes in temperature, pH and Eh, which are also very common to coal mine hydrologic regimes. Fluorescent dyes tend to decompose when exposed to sun light. The adsorption or chemical alteration of fluorescent dyes by mine environments or water chemistry can cause false negative determinations. For acid mine drainage fluorescein may be the best choice of fluorescent dyes. However, experience indicates that rhodamine WT may be less susceptible to adsorption in mining environments.

- Fluorescent dyes are contained in commonly used products such as detergents and paper as optical brighteners and antifreeze (fluorescein). These common uses often cause these dyes to be released into the environment and interfere with dye tracing possibly causing false positives. Rigorous background sampling is vitally important.

Execution of a successful tracer test requires adherence to a series of steps, which ensure that the test can answer the questions posed, and that the tracer(s) themselves are appropriate. The level of detail required for each of the steps discussed depends on the goals of the tracer test itself; however, none of the steps can be completely omitted. The steps can be generally stated as follows:

- Create a conceptual site model of the flow system using existing hydrogeological and geochemical data. Based on the model. The model should describe directions of groundwater moving, flow rates along the flow paths, and potential flow paths and discharge points.
- Define tracer test objectives. Test objectives should be defined as specifically as possible, in order to avoid ambiguity in test design. The objectives impact many other aspects of a tracer test, including the analysis methods used, sampling schedule required, number of tracers used and their properties. The extent to which test objectives can be defined, and the degree to which those objectives are accommodated in design, deployment, and analysis has a direct impact on the likelihood of success.
- Tracer selection and background testing. This step also includes defining relevant tracer properties based on the test goals. Having defined the required properties, tests are conducted to establish that the tracer candidate has those properties under expected deployment conditions. Background concentrations of tracer should be collected to ensure enough tracer is put into the system that it will be clearly discernible when it comes out.
- Tracer test design. This includes an explicit description of how the test will be conducted. The quantity of tracer, injection locations, monitoring locations, sampling method and frequency, sample analysis method, and monitoring length will be defined in the design.
- Field implementation. This is implementation of the test design, including the required documentation and variations from the test design. If the tracer is injected into a monitoring well, it is recommended to follow (chase) it with additional water to help force it into the formation. If the tracer is injected into an underground or surface mine, the additional chase water will act to push the tracer out of the well bore and into the mine pool or spoil, respectively. Once the tracer arrives at a sampling point, sample frequency may be increased to produce a complete breakthrough curve.
- Test interpretation. Test interpretation can be quantitative, qualitative, and numerical analysis, or any combination, of the test.

There are a variety of techniques to assess for the presence of tracer including grab or composite sampling, dye traps, or continuous monitoring (e.g., ion probes). Which one you use is predicated on the type of tracer and the nature of your test. A series of discrete grab samples collected on a predefined systematic interval is frequently employed. Composite samples taken over a period are less often used. Small aliquots are collected at predetermined intervals and these samples are combined to form a composite sample. There is commercially available remote sampling equipment that can collect grab or composite samples at preprogrammed time intervals. Table 2.10 summarizes three groups of methods for sampling and monitoring arrival of the injected tracers.

Visualization of fluorescent dyes at monitoring points is the most direct result of a tracer test. The most common method to sample for the tracer is to use dye traps or receptors. More than one dye can be used during the same test to improve the likelihood of a successful operation. The traps are flow-through packets of activated charcoal which will adsorb all of the fluorescent dyes available. These traps are generally about the size of a conventional teabag. The dye traps are elutriated with an alcohol solution to remove the dye. A spectrofluorophotometer is used to determine the presence and concentration of the dye in the elutriate. The recent development of relatively smaller field-deployable spectrofluorophotometers and fiber-optic probes allow direct recording of readings to data loggers on a systematic or nearly continuous basis.

Table 2.10 Description of dye monitoring methods

Sample method	Goal	Typical sample interval	Analytical method
Discrete water samples either manually or by means of automatic samplers	Quantitative tracer test by establishing dye breakthrough curve in which tracer concentrations are plotted versus time	Sampling intervals vary from a few minutes to several hours and are adjusted as the test progresses	Water samples are analyzed with field fluorometer or scanning spectrofluorometer or luminescence spectrometer in a specialty laboratory
Charcoal samplers consisting of granular activated carbon packets (often referred to as dye bugs)	Qualitative or semi-quantitative tracer test by determining presence of tracer in monitoring locations. Breakthrough curves can be plotted but the concentrations are not quantitative	Charcoal samplers are deployed at intervals varying from 2 to 14 days and are adjusted as the test progresses	Retrieved charcoal samplers are eluted in a specialty laboratory, and the eluted samples are then analyzed by scanning spectrofluorometer or luminescence spectrometer
In-situ continuous monitoring	Screening level tracer test with flow-through fluorimeter	High frequency sampling such as every 5 min can be accomplished	Use downhole or flow-through in-situ dye fluorometry or dye sensors

Ion probes can be installed at points where chemical tracers are anticipated to show up. These probes can be read manually on predetermined intervals or interfaced with automatic data loggers set to record at the desired intervals. The data loggers are periodically downloaded to computers for subsequent analysis.

The primary result of a quantitative tracer test is the breakthrough curve in which tracer concentrations are plotted versus time. A well-documented breakthrough curve is more convincing evidence for an underground connection than an isolated result from a single charcoal bag or from one or two water samples. Furthermore, several relevant flow and transport parameters can be directly obtained from the breakthrough curve, such as the time of first arrival, the peak time, and the maximum concentration. Concentrations are either plotted as absolute values or normalized by the respective injection quantity to allow better comparison of breakthrough curves from multi-tracer tests.

The peak concentration of a tracer breakthrough graph represents the average ground water velocity. The portion of the concentration that increases above background observed prior to and after the peak (spreading out of the tracer concentration) is caused by hydrodynamic dispersion. Hydrodynamic dispersion is a combination of mechanical dispersion (mixing) and molecular diffusion of the tracer. Molecular diffusion is the spreading of the tracer driven mainly by its thermal kinetic energy in the direction of the concentration gradient.

Sampling beyond the peak will allow for estimations as to how much of the tracer that was put into the system came out, which yields information such as flushing rates in the case of underground mines and provides insight into the ground-water regime of surface mine backfills. The area under the graph along with the discharge rate will allow for mass loading calculations.

2.3.4 Physical Model Simulations

The occurrence of mine subsidence is typically affected by multiple factors, and its processes are not readily amenable to mathematical expressions or numerical modeling. Instead, physical models can be constructed to investigate subsidence mechanisms and perform root cause analysis for subsidence hazards. Figure 2.6 shows a physical model of 2 m long, 0.3 m wide, and 1.8 m high (Zeng et al. 2023). The upper part is composed of unconsolidated soil simulating a porous-medium aquifer, while the lower part is made of materials simulating the bedrock overlying the coal seam. The mining thickness is 3.5 cm. The model simulates fully mechanized longwall mining. The stress in MPa is measured at regular intervals of approximately 20 cm. During the advancement of the working face, the breaking of the overlying rock structure, stress changes and water flow conditions were observed and recorded.

When the working face advanced from 8 cm to 38 m, there is no obvious change in the floor stress and the monitored water pressure, which is relatively regular. When the working face advanced from approximately 103 to 113 cm, the overlying rock was seriously damaged, and the fractures grew upward and expanded lateral. When the

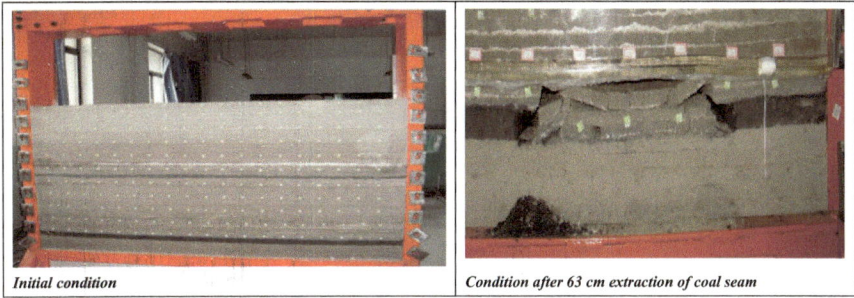

| Initial condition | Condition after 63 cm extraction of coal seam |

Fig. 2.6 An example of physical model and simulation

working face advanced to 123 cm, induced fractures connected with the overlying soil, and water inflow first appeared. As the fracture development of the bedrock continued to grow, water flow rate increases, and surface began to deform.

Successful application of the physical model relies on representativeness of the collapse conceptual site model at the site. The physical model constructed for a project site can be surgically excavated after simulations to visualize the soil deformation inside the soil. Some marker layers can be constructed with the soil to facilitate visualization. Like numerical models, the results from the physical model simulations need to be interpreted with caution because of limitations in scale and boundary conditions of the physical model.

2.4 Subsidence Monitoring

Subsidence monitoring is essential to understand the dynamics of subsidence development and predict the impact of subsidence. Monitoring shall be implemented before, during, and after mining operations. The objectives of subsidence monitoring are designed to:

- Collect data for continued understanding and validation of mining related impacts and refinement of the conceptual site model of subsidence
- Develop threshold and alarm levels for early warning and detection of subsidence impacts before surface impacts occur
- Identify surface movements due to mining
- Implement corrective actions and contingency plans if needed.

A monitoring strategy utilizes an integrated approach consisting of ground movement monitoring, subsurface soil erosion and deformation monitoring, and dynamic condition monitoring. An effective subsidence monitoring program is to help with the following:

- Collection of extensive data before mining starts to establish robust baseline and background data for the area above and adjacent to the mine.
- Sequence of mining with a starting point far away from key areas of concern or features (infrastructures, residences, roadways, or surface water bodies). This arrangement will allow time to gather technical information and inform the extent of subsidence.
- Design of a sufficient rock barrier as the buffer zone between the fractured zone and key areas of concern even after mining has ceased.
- Monitor the effect of mining on key features at key positions relating to the mining front. The monitoring effort is often based on subsidence predictions from modeling exercises. The early monitoring data shall be collected far in advance of subsidence movement to protect the key features.
- Periodically assess, analyze, and interpret monitoring data to demonstrate compliance with pertinent regulations and identify substantive variations from predictions. Depending on the importance of the key features to be exposed to the subsidence risk and complexity of the subsidence processes, monitoring reports may include daily reports, weekly reports, monthly reports, quarterly reports, or annual reports.

2.4.1 Monitoring Methods and Instrumentation

A successful monitoring system provides the means to measure ground deformation, grade changes, displacement, and anticipate and assess the potential for collapses. Different methods and instrumentation types using proven technology and best practice has been used to monitor surface subsidence. As such, methods and instrumentation will improve over time. The monitoring methods and frequencies of measurement recommended represent a best estimate on how caving, fracturing, and the associated subsidence will progress. When mining commences and data starts becoming available, adjustments should be made on the monitoring methods or frequency based on actual subsidence behavior to ensure that the data is fully representative of conditions.

Table 2.11 presents a list of commonly used monitoring techniques and the primary purpose and encompasses the subsidence monitoring window period from pre to post mining activities. Selection of monitoring methods for a specific site depends on the mine subsidence conceptual site model, monitoring level, and resources to continue monitoring through time.

Some of the monitoring methods (geophysical techniques and techniques used in boreholes) are to monitor the overburden deformation in the subsurface, whereas other methods (aerial techniques) are to monitor the surface deformations. Surface deformation monitoring is relatively straightforward (Jones and Blom 2015), whereas the subsurface deformation monitoring is more complicated. The methods that are used to monitor overburden deformations should consider the site-specific conditions, the project objectives, resources available, and concealment and suddenness.

Table 2.11 Subsidence monitoring techniques

Monitoring method	Purpose
InSAR (interferometric synthetic aperture radar)	Measure changes in surface displacements (primary vertical sense with lateral calibration to other instruments) across a very large area
Aerial photogrammetry (drone technology)	Measure vertical and lateral (using targets in the field) displacements through 3-dimensional point cloud analysis of the surface. It also provides high-definition images for observational analysis of subsidence behavior. Digital terrain models are typically generated from aerial photography to calculate the relative displacements of the surface ground
Robotic prism network using total survey stations	Monitoring of survey prisms anchored to the surface at strategic points to validate the relative displacements captured through InSAR and photogrammetry analyses
LIDAR (light detection and ranging)	Terrestrial laser scanning of key features or areas of concern for toppling/rotational type of displacements
GPS (global positioning system) survey	Installed in selected locations around the proposed subsidence area. The three-dimensional location is surveyed automatically to provide the relative displacements of the ground
TDR (time domain reflectometers) and BOTDR (Brillouin optical time domain reflectometry)	Monitor shallow to deep seated subsurface rock mass deformations and assist in tracking of caving or fracturing propagation to surface during mining. The cables are installed and grouted into a borehole or trenches near the surface for monitoring ground deformations and movements. TDR/BOTDR are limited in that they can only provide area/location of ground movement and not magnitude
Video camera in open holes	Track and monitor the propagation of the cave to surface using simple borehole camera and/or weighted probe
Smart markers and beacons with detectors	Monitor caving/fracturing growth, propagation and flow of material in the mined-out areas
Wireless in-ground monitoring ("Geo4Sight")	Monitor shallow to deep seated subsurface rock mass deformation with magnitude and direction ability
Inclinometers or tilt meters	They measure the angle of movement from the horizontal position in microradians (μrad) and deliver the data collected to the onsite monitoring computer that is part of integrated early warning system. An example is the horizontal in-place inclinometer strings, which are a type of tiltmeter composed of micro-electro-mechanical sensors encased in a sealed housing. The strings can be used for early warning of subsidence near infrastructure with an accuracy of $\pm$ 0.006 in. per 10 ft., which gives a detailed ground profile along the array extent. May not tolerate high levels of rock mass deformation

(continued)

Table 2.11 (continued)

Monitoring method	Purpose
Extensometers	These are electronic cable devices that are installed and anchored in the bottom of a borehole casing with a recorder to monitor lateral displacements of overburden materials. This type of instrumentation can detect changes in land surface elevation to 1/100th of a foot, particularly useful close to areas of concern/importance as they are very sensitive and provide a high degree of accuracy
Surface monuments (concrete lock blocks)	Monitor vertical, but primarily the lateral surface displacements (direction and magnitude) for calculating the angular distortion for any selected area
Geophysical detection techniques (ground penetrating radar or earth resistivity imaging)	They collect data that is associated with properties of earth materials. Time-lapse surveys track changes of the formation layers in the subsurface. Viable techniques to identify anomalies in the subsurface before subsidence appears on surface
Microseismic monitoring system	An array of seismic sensors and geophones installed in the ground or monitoring wells to detect and determine the location and magnitude of potential seismic events
Crack mapping	Field observations of subsidence induced tension crack formation. Survey measurements of location, length, aperture and vertical offset. Observations can be done initially using the photogrammetry images to identify onset and followed by field measurements and installation of "crackmeters" to monitor growth and trends

In a project that involved a brine cavern with the potential to catastrophically fail (Fergason et al. 2015), several monitoring techniques including microseismic arrays and borehole tiltmeters were used, and the borehole tiltmeter application was considered to be one of the most effective tools in the early warning monitoring system for collapses. In applications to linear engineering projects such as pipelines, roadways, or high-speed rail roads, TDR and BOTDR have been demonstrated to be viable techniques to identify subsidence risk areas (Guan et al. 2015; Jiang et al. 2006; O'Connor et al. 2001).

2.4.2 Subsidence Monitoring Plan

A subsidence monitoring work plan includes the following monitoring activities:

- Layout monitoring points for surface and subsurface deformations
- Equipment and instrument verification and calibration
- Monitoring equipment installation, inspection, and verification
- Instrument monitoring schedule
- Collect data and supplement on-site investigations

- On-site monitoring, monitoring data processing, analysis, and information feed-back
- Submit monitoring data and summary report.

The level of effort in subsidence monitoring depends on the importance of the key features or areas of concern, their sensitivity to subsidence, distances from mining, and mining phases. Figure 2.7a shows a hypothetical mining site to illustrate the requirements for monitoring techniques and frequencies. For discussion purposes, four key areas (features) of concern, i.e., a historically important site, a surface river, a roadway, and a residential area, are included in this figure. Also shown in the figure are the lateral extents of subsidence that were predicted from a finite-difference Fast Lagrangian Analysis of Continua program (FLAC3D) for three timeframes—10, 20, and 40 years after mining starts. The mine is assumed to operate for 40 years. Figure 2.7b shows the vertical profiles of the mine site and the vertical extents of subsidence predicted from the modeling exercises. Several zones are depicted in the figure and explained as follows:

- The subsided zone is the area where subsidence is visible and usually located just above the mined-out area or goafs. This zone is characterized by the greatest vertical displacement.
- The fractured zone is the area outside the subsided zone where visible deformation (cracking and dislocations) can be seen. The fractured zone is characterized by radial cracks and typically rotational and toppling mode failures.
- The area outside of the fractured zone is deemed the continuous subsidence zone. This zone is characterized by extremely small rock deformations that can only be detected using high-resolution monitoring equipment such as extensometers, and tilt meters. If deformations are significant enough, in some cases they can create small hairline cracks in the surface of concrete but will not be visible in the soil or on the ground. This zone is also commonly referred to as the elastic zone, because the deformations are usually below the damage threshold for rocks.
- The zone adjacent to the continuous subsidence zone is the stable zone. The stable zone can be characterized as the zone where rock is essentially undisturbed.

Table 2.12 summarizes monitoring techniques and monitoring frequencies. It is important to reiterate that the suggested monitoring techniques and frequencies are for illustrative purposes only. Researchers need to design their own monitoring program based on site-specific conditions and needs.

Pre-mining monitoring and baseline studies: This phase establishes baseline data readings of the entire area encompassing all the features/areas of particular interest. This is to establish whether there are any observable displacements that might be occurring in the absence of mining and if there are any seasonal fluctuations that are evident. It also allows one to establish fixed ground control targets and system calibration for future data processing and monitoring. Key features or areas of concern above planned underground mines can be monitored using terrestrial LIDAR scans, InSAR and select rock spires using digital tilt meters. This baseline data is collected

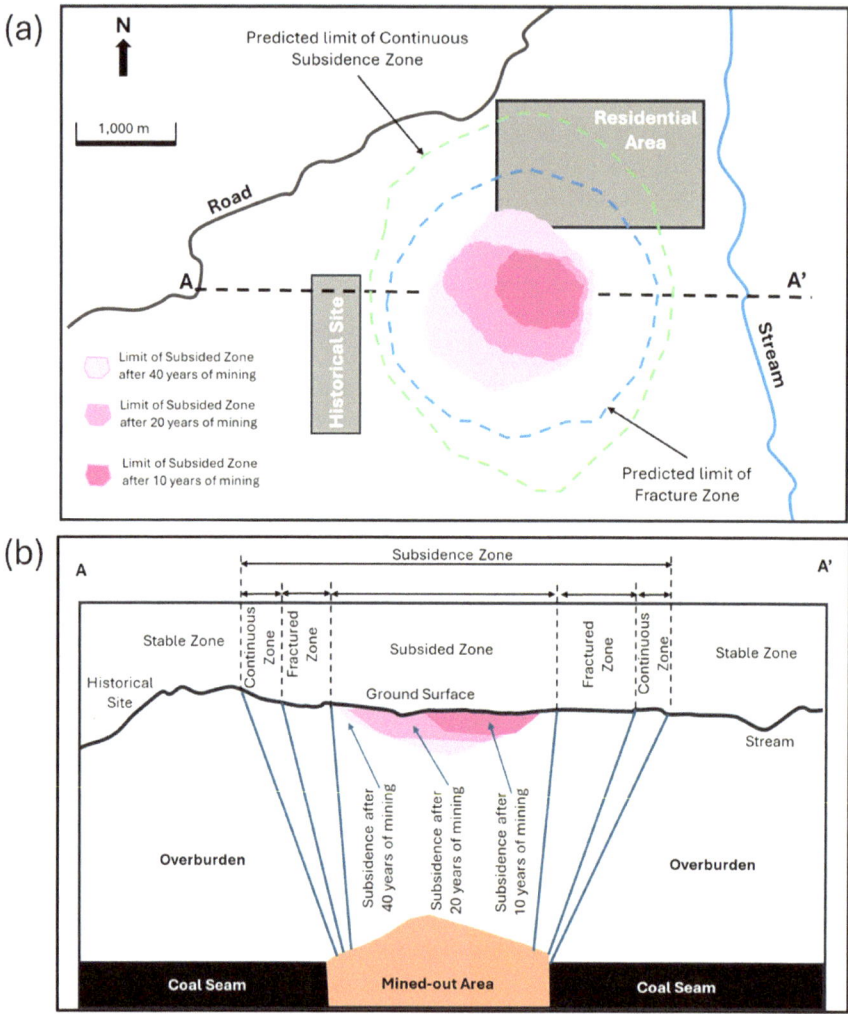

Fig. 2.7 Schematic conceptual model of mining-induced subsidence in a hypothetical mine

using periodic (e.g., biannual) LIDAR scans and InSAR satellite image of the surface and rock formations for future comparisons before, during and after mining. Experience from the use of tilt meters shows that it is feasible to accurately and remotely monitor pillar deformation on a nearly continual basis.

Monitoring during mining: During this phase underground mining has been initiated through production, and the mining-induced caving and fracturing start to propagate toward the surface. The surface area predicted to be affected by mining should be restricted from public access. The surface area tends to expand with mining progress.

Table 2.12 Subsidence monitoring schedule for each key feature or area of concern

Monitoring period	Historical site		Stream		Road		Residential area	
	Monitoring technique	Frequency	Monitoring technique	Frequency	Monitoring technique	Frequency	Monitoring technique	Frequency
Pre-mining baseline monitoring	• Aerial surveys • LIDAR scans • Google Earth observation	Twice a year	Surface joint mapping	Once prior to mining	• Observation of road conditions • Google Earth observation • Survey marks • GPS or prisms	Once prior to mining	• Aerial surveys • Google Earth observation	Twice a year
							• Survey markers • GPS or prisms • Extensometers	Once prior to mining
	Tiltmeters installed on selected pillars	Continuously	Occurrence surveys and measurement of base flow	Quarterly			Crack displacement monitors	Once prior to mining
							Tiltmeters installed on selected pillars	Continuously
			Measurement of flow in streams and water captured in mine	Quarterly and after precipitation events				
Monitoring during mining	• Aerial surveys • LIDAR scans • Prisms survey • Extensometers • Google Earth observation	Annually	Surface joint mapping	Once a year	• Observation of road conditions • Google Earth observation	Once every quarter	• Survey marks • GPS or prisms • Extensometers	Once a year
	InSAR imaging	Twice a year					• Aerial surveys • Google Earth observation • Crack displacement monitoring	Monthly

(continued)

Table 2.12 (continued)

Monitoring period	Historical site		Stream		Road		Residential area	
	Monitoring technique	Frequency	Monitoring technique	Frequency	Monitoring technique	Frequency	Monitoring technique	Frequency
	Tiltmeters installed on selected pillars	Continuously	Subsidence fracture mapping	Once a year			• Smart markers/caving trackers • TDR installed before caving initiation	Continuously
	Micro-seismic monitoring	Continuously					Micro-seismic monitoring	Continuously
Post-mining monitoring	• Aerial surveys • LIDAR scans • Google Earth observation	Annually for 15 years after mining stops	Occurrence surveys and measurement of base flow	Quarterly and after precipitation events	• Observation of road conditions • Google Earth observation • Survey marks • GPS or prisms	Once every 2 years for 15 years after mining	• Smart markers/caving trackers • TDR installed before caving initiation	Once every 2 years for 15 years after mining
	Seismic monitoring	Once every two years for 15 years after mining stops	Surface joints mapping	Once every 2 years for 15 years after mining stops			Tiltmeters installed on selected pillars	Continuously
	Tiltmeters installed on selected pillars	Continuously for 15 years after mining stops					INSAR imaging	Continuously
							Micro-seismic monitoring	Continuously

Surface subsidence can be monitored through the use of available industry best practice and demonstrated technology including, but not limited to extensometers, GPS or survey prisms, crack displacement monitoring, TDR cables, aerial photography; InSAR, microseismic monitoring system, and smart markers and caving trackers.

Post-mining monitoring: This phase is characterized by time dependent compaction of the broken material within the mining-induced disturbance zones. This compaction of material results in residual subsidence or "creep" which can occur over several years in the former mobilized zone, however the incremental changes or rate of subsidence tends to reduce rapidly in an exponential fashion. Examples and case studies of residual subsidence show that it can decay with good certainty within approximately 15 years, which is a conservative estimation as long wall mining. Manmade rockfill studies have a much shorter duration for residual subsidence (up to 10 years). As such, monitoring the impact of surface subsidence on key infrastructures should continue for at least 15 years after mining using the same techniques that are used during mining.

2.4.3 Subsidence Trigger Action Response Plan

The subsidence monitoring plan and the data obtained from the various instrumentation are fundamental input sources in the overall management of subsidence. Tracking against predicted states of subsidence provides insights into trends and allows for projections and implementation of early mitigation measures, if required. The trigger action response plan is to ensure that a timely and planned action is ready to mitigate the risk when certain triggers are reached or detected by the monitoring program. The trigger level is technique specific and can be established for each monitoring technique. Alternatively, a weight of evidence approach can be used to integrate data from various monitoring methods. For simplicity and purpose of early warning, the surface features are either sensitive to tilts or to angular distortion. Tilt is the derivative of the vertical displacement with respect to the horizontal displacement. Tilt occurs when a land surface has different levels of vertical subsidence on either end. Surface features most affected by tilt are typically tall structures and structures such as steep slopes.

- Under normal conditions, the average tilt values should be less than $2°$.
- When the tilt values are between 2 and $4°$, early warning may be issued to pay special attention during field reconnaissance and to increase the monitoring frequency.
- When the tilts values exceed $5°$, corrective actions should be taken to avoid damage to the key features or areas of concern.

Angular distortion is the ratio of the differential settlement between two points divided by the distance between them. The mining-induced subsidence naturally creates the differential imbalance resulting in the body distortion which in turn generates various strain gradients with tension cracks and scarps ultimately appearing as surface expressions. Table 2.13 presents the general guidelines on the trigger levels for the hypothetical site discussed above. These guidelines serve as a tool to manage the subsidence risks and prevent damage to the critical surface infrastructures.

Based on the deviation tolerances from the predicted results, the example assigns three trigger levels:

- Level 1 ≤ 150 m
- Level 2 ≥ 150 m and ≤ 250 m
- Level 3 ≥ 250 m.

The thresholds established for the trigger levels assume that the prediction model represents reasonably well the actual mining conditions, and no significant impacts are identified on key areas of concern from the predictions. The modeling results may be updated as more data becomes available with the mining progress. The actions associated with each level are commensurate to the level condition, with Trigger Level 1 being the lowest and Level 3 requiring more stringent mitigations. Early warning is to be issued at Trigger Level 1, whereas correction actions are to be taken at Trigger Level 3. If Trigger 3 is activated and the situation is not investigated or appropriate remedial/mitigation measures are not put in place, the offset will be retained through later stages of mine operation, and the areas of concern may be impacted as the subsidence bowl increases with progressive subsidence growth.

Figure 2.8 shows an example to illustrate how the proposed trigger levels are applied to subsidence monitoring at the 20th year of mining. The measured results in Example fall below the modelled condition and therefore no actions are needed, and mining continues as planned. The actual measurements in Example 2 show that the angular distortion values are beyond the predicted range, but the offset is less than 150 m, which meets the requirements for Trigger Level 1. The actual measurements in Example 3 show that the angular distortion values are beyond the predicted range, while the offset is between 150 and 250 m, which meets the requirements for Trigger Level 2. The actual measurements in Example 4 show that the angular distortion values are beyond the predicted range with the offset exceeding 250 m, which meets the requirements for Trigger Level 3. The associated actions in response to Levels 1, 2 and 3 are summarized in Table 2.13.

Table 2.13 Subsidence trigger levels and corresponding action responses

Trigger level	Threshold and condition description	Action/remediation measures
No trigger activated	Actual subsidence plotted against the modelled is within the prediction state	• Mining continues as planned • Subsidence monitoring continues as planned
Level 1	Actual subsidence plotted against the modelled is in excess of the prediction state with an offset not exceeding 150 m	• Confirm the instruments all function correctly and provide data on the intervals as required • Conduct an internal data validation exercise to check on the data integrity and interpretations • Inform the mine management of the outcome following the interrogation of the data • Inform regulatory agencies of the subsidence exceedance condition, as appropriate • Conduct a field investigation to assess and document the surface conditions specifically in the zone where the subsidence exceedance has occurred • Increase monitoring frequency and continue to track for any further deviations
Level 2	Actual subsidence plotted against the modelled is in excess of the prediction state with an offset of 150–250 m	• Confirm the instruments all function correctly and provide data on the intervals as required • Conduct an internal data validation exercise to check on the data integrity and interpretations • Engage with an external and independent person/ organization to conduct a similar instrumentation and data integrity review • Inform the mine management of the outcome following the interrogation of the data • Inform regulatory agencies of the subsidence exceedance condition, as appropriate • Conduct a field investigation to assess and document the surface conditions specifically in the zone where the subsidence exceedance has occurred • Alter the mining sequence and plan schedule to reduce potential further expansion or exceedance of the footprint by targeting production that would locate furthest from the surface feature of concern • Add instrumentation (to fill any possible data gaps or improve sensitivity) and further increase monitoring frequency and continue to track for any further deviations

(continued)

Table 2.13 (continued)

Trigger level	Threshold and condition description	Action/remediation measures
Level 3	Actual subsidence plotted against the modelled is in excess of the prediction state with an offset of greater than 250 m	• Immediately confirm the instruments are all functioning correctly • Conduct an immediate data validation exercise to check on data integrity • Immediately inform the mine management on the condition observed • Notify the regulatory agencies of the situation immediately after having conducted the internal data validation • All mining activities cease immediately • Investigation to be initiated immediately to thoroughly review the condition and identify possible causes for the exceedance level in the subsidence profile • Implement remedial measures as required and determine the options if restarting the mining production is possible at this time

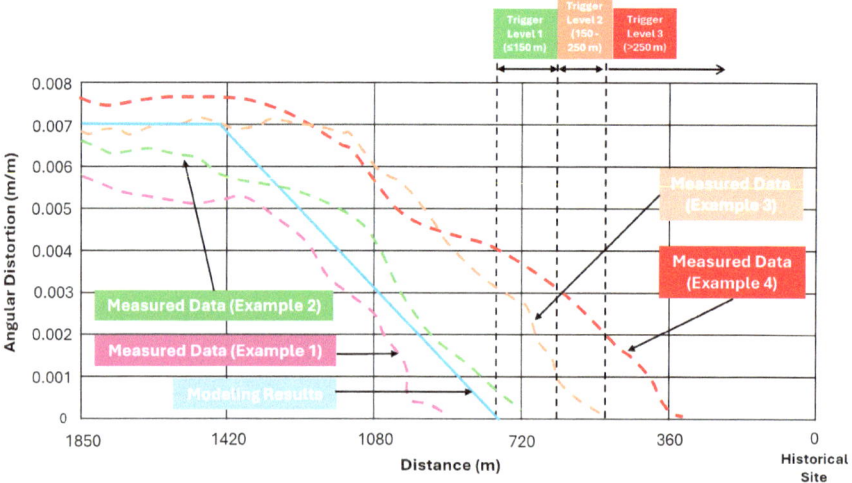

Fig. 2.8 Application of trigger levels to subsidence monitoring (at 20th year of mining)

References

Bechtel TD, Bosch FP, Gurk M (2007) Geophysical methods. In: Goldscheider N, Drew D (eds) Methods in karst hydrogeology. International contribution to hydrogeology, IAH, vol 26. Taylor and Francis/Balkema, London, pp 171–199

Erchul RA, Slifer DW (1987) The use of spontaneous potential in the detection of groundwater flow patterns and flow rate in karst areas. In: Beck BF (ed) 2nd multidisciplinary conference on sinkholes and the environmental impacts of karst, Orlando, Florida, pp 217–226

Fergason KC, Rucker ML, Panda BB (2015) Methods for monitoring land subsidence and earth fissures in the Western USA. Proc IAHS 372:361–366

Guan ZD, Jiang XZ, Wu YB, Pang ZY (2015) Study on monitoring and early warning of karst collapse based on BOTDR technique. In: Doctor DH, Land L, Stephenson JB (eds) Sinkholes and the engineering and environmental impacts of karst: proceedings of the fourteenth multidisciplinary conference, NCKRI symposium 5, pp 407–414

Jiang XZ, Lei MT, Chen Y (2006) An experiment study of monitoring sinkhole collapse by using BOTDR optical fiber sensing technique. Hydrogeol Eng Geol 33(6):75–79

Jones CE, Blom RG (2015) Pre-event and post-formation ground movement associated with the Bayou Corne sinkhole. In: Doctor DH, Land L, Stephenson JB (eds) Sinkholes and the engineering and environmental impacts of karst: proceedings of the fourteenth multidisciplinary conference, NCKRI symposium 5, Rochester, Minnesota, pp 415–412

Lange AL, Barner WL (1995) Application of the natural electrical field for detecting karst conduits on Guam. In: Beck BF (ed) Karst geohazards. Balkema, Rotterdam, pp 425–441

O'Connor K, Clark R, Whitlatch D, Dowding C (2001) Real-time monitoring of subsidence along I-70 in Washington, Pennsylvania. Transp Res Rec J Transp Res Board 1772(1):32–39

Sheets RA (2014) Use of electrical resistivity to detect underground mine voids in Ohio. Water-resources investigations report 02-4041. U.S. Geological Survey

Woods DA, Marinello SA (1999) Solution void cavity definition using the surface based Z-scan high resolution magnetotelluric system. Solution Mining Research Institute Spring Meeting, Las Vegas, Nevada

Zeng YF, Pang ZZ, Wu Q, Lian HQ, Du X (2023) Roof water disaster in coal mining in ecologically fragile mining areas: formation mechanism and prevention and control measures. Springer, Cham. https://doi.org/10.1007/978-3-031-33140-4

Zhou W, Beck BF, Stephenson JB (1999) Investigation of groundwater flow in karst areas using component separation of natural potential measurements. Environ Geol 37(1–2):19–25

Zhou W, Beck BF, Adams AL (2002) Effective electrode array in mapping karst hazards in electrical resistivity tomography. Environ Geol 42:922–928. https://doi.org/10.1007/s00254-002-0594-z

Chapter 3
Remediation of Goaf Areas for Resettlement in Yulin, Shaanxi Province, China

Abstract Mining of coal seams in the study area has created single-layer goafs of approximately 21,079 m^2. The underlying goafs of are in an unstable state, and subsidence or collapses may occur if no engineering measures are taken. The presence of the goafs has become one of the key factors restricting the land use of the area. The goafs are remediated to build a residential community for resettlement of residents. Three types of boreholes are drilled to remediate the potential subsidence area: hydrogeological investigation boreholes, grout curtain boreholes, and grouting boreholes. Two hundred fifty-two boreholes are completed with drilling footage of approximately 12,067 m. Approximately 55,327 m^3 of grout were used to address the coal seam goafs.

Keywords Mining-induced goaf · Grout curtain · Grouting boreholes · Land redevelopment · Remediation

3.1 Introduction

The project area is known for rich coal resources, and coal mines are widely distributed. Mining of coal seams 4^{-3} and 5^{-2} has created single-layer goafs on the east and west sides of the site. The underlying goafs of the proposed site are in an unstable state, and subsidence or collapses may occur if no engineering measures are taken. The presence of the goafs has become one of the key factors restricting the land use of the area. The goafs are remediated to build a residential community for resettlement of residents.

The total area of the goafs is approximately 21,079 m^2, of which 7741 m^2 is associated with mining of the 4^{-3} coal seam, and 13,338 m^2 is associated with mining of the 5^{-2} coal seam. Three types of boreholes are drilled to remediate the potential subsidence area: hydrogeological investigation boreholes, grout curtain boreholes, and grouting boreholes. Two hundred fifty-two boreholes are completed with drilling footage of approximately 12,067 m. The work plan requires a grouting volume of 10,416 m^3 to address the 4^{-3} coal seam goaf and 25,798 m^3 to address the

5^{-2} coal seam goaf, respectively, totaling 36,214 m^3. The actual grouting volume is approximately 55,327 m^3.

3.2 Main Remediation Components

The main remediation components include drilling in goaf areas, grouting and sand filling in goaf areas, construction of water storage tanks, mixing tanks, cement fly ash tank foundations, water quality monitoring, post-remediation performance testing and monitoring. The following works are completed:

Goaf Drilling

- Number of borehole: 252 boreholes
- Footage: 12,067 m
- Borehole specification: 150 mm opening, drilled to 5 m below stable bedrock in 4^{-3} coal seam goaf, drilled to 5 m below muddy sandstone in 5^{-2} coal seam goaf. Then changed to 91 mm to 1 m below the coal seam floor
- Orifice pipe: 127 mm steel pipe, wall thickness 4 mm, down to the change of diameter position.

Grouting and Sand Filling in Goaf

- Total volume of grouting and sand filling in goaf: 54,280 m^3
- Volume of goaf grouting: 33,846 m^3
- Volume of sand filling of goaf: 20,434 m^3
- Slurry materials: water, cement, fly ash, water glass, aeolian sand
- Slurry ratio: water-cement ratio is 1:1–1:1.3, cement and fly ash ratio is 2:8–3:7
- Aeolian sand filling volume: 20% of the total filling volume
- Water glass content: 2% cement weight.

Water and Mixing Tanks

- Number of water tanks: 2
- Masonry material of water tank: red brick
- Masonry thickness: 240 mm
- Dimension: length 6 m × width 4 m × depth 2 m
- Number of mixing tanks: 6
- Masonry material of mixing tanks: red brick; masonry thickness: 240 mm
- Dimension: diameter 2.5 m, depth 3.0 m
- Surfacing: 16 mm thick, including 6 mm 1:0.25 cement mortar and 10 mm 1:0.3 cement mortar.

Foundation of Cement Fly Ash Tank

- Foundation size: 2 m wide, 3 m long, and 2.5 m deep
- C20 concrete pouring.

Water Quality Monitoring

- Water level, color, pH value, turbidity during the construction period
- Complete water quality analysis.

Post-construction Performance Monitoring

- Number of performance monitoring borehole: 1
- Drilling specifications: 150 mm opening to 5 m below bedrock, then diameter changes to 91 mm to drill to 1 m below the floor of the goaf
- Tested for wave velocity and televiewing inside the inspection borehole
- Uniaxial compressive strength test of the core samples
- Monitoring parameter: horizontal and vertical displacement
- Monitoring level: second class
- Complexity classification: complex.

3.3 Remediation Overview

3.3.1 Remediation Environment

There are two mineable coal seams, namely the 4^{-3} coal seam and 5^{-2} coal seam. The characteristics of each coal seam are as follows:

- Coal seam 4^{-3} is located at the top of the second upper cycle of the Yan'an Formation, with simple layered beds and stable stratigraphic distribution. The stratum underlying 4^{-4} coal seam is relatively stable, ranging from 9 to 26 m, with an average of 14 m. The occurrence is controlled by regional structural and paleotopographic conditions. It is a monocline layer tilted to the northwest. The later structural influence is small, and the dip angle is less than $1°$. This coal seam is distributed throughout the region, and the mining area is distributed in the northwest. The mining area is about 1.87 km^2, the coal seam thickness varies from 1 to 3 m. The mining thickness varies from 1.8 to 3.0 m, and the average thickness is 2.1 m. It does not contain interlayers, has a simple structure, and has an obvious thickness variation pattern.
- 5^{-2} coal seam is located at the top of the lower cycle of the first section of the Yan'an Formation. It is produced in a simple layered manner. There is no obvious expansion and contraction mutation phenomenon, and the stratigraphic position is stable. The occurrence is controlled by regional structural and paleotopographic conditions. According to the bottom plate contour, except for the wide and gentle undulations in the northeast, the overall monocline is tilted to the northwest. The later structural influence is small, and the dip angle is less than $1°$.

This coal seam is the main mining coal seam in the area, with a mining thickness ranging from 2.19 to 6.12 m, an average thickness of 3.89 m, and a clear thickness variation pattern. The coal seam generally contains one layer of interlayer gangue, and two layers in some areas, with a simple structure. The thickness of the interlayer gangue is 0.03–0.52 m, and the lithology is mainly carbonaceous mudstone and mudstone, with a small amount of fine sandstone. The coal seam has a stable position, a clear thickness variation pattern, and the coal type is long flame coal. The structure is simple, and it is a stable thick coal seam that can be mined in the entire area.

The lithology of the direct roof and floor of the coal seam is mainly mudstone. According to the on-site grouting construction, it was found that the project area is the final mining area, and the recovery rate is much higher than the average mining level of room-pillar mining, and it is as high as 80% in some areas. The connectivity of the curtain boundary goaf area is good, and the curtain grouting dissipation is large; at the same time, the goaf area is shallowly buried, the stratum joints and fissures are developed, and the overburden permeability is good.

3.3.2 Remediation Technology

3.3.2.1 Borehole Placement

Remediation design mainly includes the layout and survey of the boreholes. According to the coordinates of the borehole positions on the design drawings, RTK is used to measure and lay out each borehole in the field, and wooden stakes and red flags are used to mark the borehole positions. The hole position deviation is less than 0.5 m. The grouting boreholes are measured and laid out on site using RTK and steel ruler. The position of the borehole should not deviate from the design position by 2.0 m. If the borehole cannot be placed at the design position due to the influence of the terrain, the nearby boreholes that can be placed will be drilled first, and then adjustments are made based on the data of the goaf revealed by the nearby drilling. Deviations from the design should be approved by the design representative, the owner and the lead engineer.

3.3.2.2 Drilling Operation

The boreholes start with an Φ150 mm opening hole. In the 4^{-3} coal seam goaf area, the boreholes are drilled to 5 m below stable bedrock and then the φ133 mm pipes are installed. In the 5^{-2} coal seam goaf area, the boreholes are drilled to 5 m below mud sandstone and then the φ133 mm pipes are installed. The pipes are made of hot-rolled seamless steel with a wall thickness of 4 mm. Full-mouth welding is used between pipes. After the pipe is stabilized, the drilling diameter is changed to 91 mm until the end of the borehole. The borehole casing extends 30 cm above the ground.

The boreholes are completed using cement slurry with a water-cement ratio of 1:1.5 (casting length 5–10 m, 2% of cement weight of accelerator should be added to the slurry). The curing time of the cement slurry is not less than 24 h. The orifice pipe should be 30 cm above the ground, and the top double-sided welded flange (inner

Fig. 3.1 Processes of drilling operations

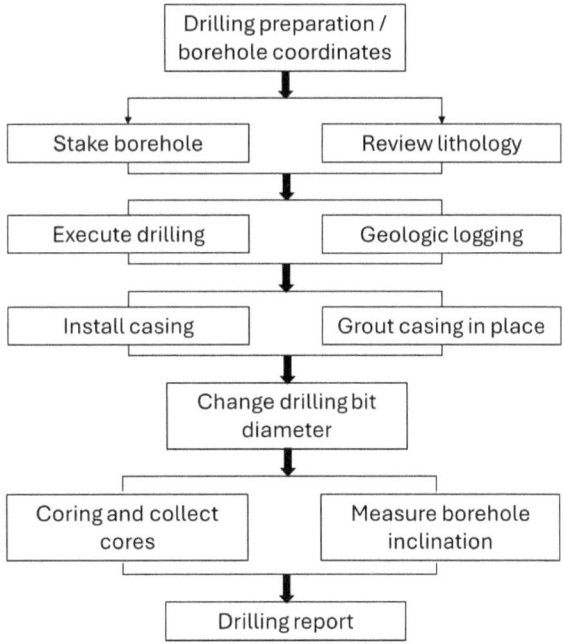

diameter 133 mm, thickness 10 mm) should be connected to the grouting pipeline. The drilling engineering process is shown in Fig. 3.1.

3.3.2.3 Grouting Materials

Grouting Materials and Requirements

Grouting materials mainly consist of water, cement, fly ash, quick-setting agent (water glass) and aeolian sand.

- Cement: Portland cement with a strength grade of 32.5R is used. Its quality complies with the national standard of "General Portland Cement" (GB175-2007/ XG2-2015). When the cement is damp or stored for more than 3 months, it should be sampled and evaluated.
- Water: Mixing water complies with the requirements of the "Concrete Water Standard" (JGJ63-2006) with a pH value greater than 4.
- Fly ash: The fly ash used for remediation is the primary waste product of the power plant (thermal power plant). Its quality meets the standard of "Fly ash used in cement and concrete" (GB/T 1596-2017).
- Aeolian sand: Local aeolian sand, which contains no debris or mud.
- Admixture (quick-setting agent): The admixture is a water glass solution with a modulus between 2.4 and 3.4. The admixture accounts for 2.0% of the cement

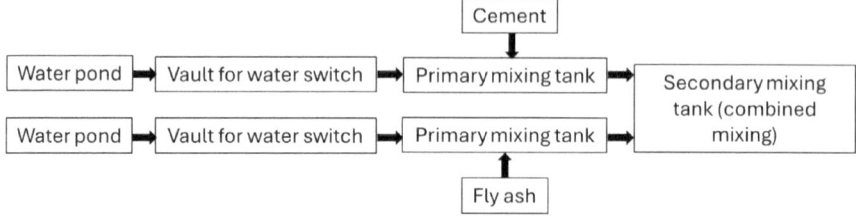

Fig. 3.2 Processes of making grouting materials

volume. Sampling inspection is required for each batch or every 30 tons. The process of making the grouting materials is shown in Fig. 3.2.

Slurry Ratio

The grouting slurry is cement fly ash slurry with the following ratios:

- Water to solid ratio: 1:1–1:3.
- Cement to fly ash ratio: 2:8–3:7.
- Aeolian sand filling volume: Approximately 20% of the total filling volume that is determined based on the on-site filling situation and the strength of the test block that meets the design requirements.
- Quick-setting agent: 2% of the weight of cement is added to the curtain borehole. When the grouting volume of the grouting hole is larger than 100 m^3, 2% of the weight of cement is added to the slurry.
- Before construction, the slurry ratio test is conducted in the laboratory according to the cement and fly ash used during production (the ratio is 1:1.1, 1:1.2, 1:1.3, and the solid ratio is 2:8 and 3:7). The testing parameters include the dry material concentration of slurry, slurry density, initial and final setting time, and consolidation rate.

Grouting System Configuration

The grouting system consists of storage tank, primary mixing tank (machine), secondary mixing tank (machine), water supply system, grouting pump, grouting pipeline, orifice pipe, orifice device, sand feeder, and sealing device.

3.3.2.4 Grouting and Sand Filling Operations

Curtain Borehole Grouting Control

Grouting of curtain boreholes is the key to control costs and ensure success. In order to make the curtain borehole grouting slurry solidify as soon as possible to form a curtain and prevent excessive loss of slurry to non-grouting parts or sections, the following measures are taken:

- Increase the sand ratio in the boundary boreholes. For boundary boreholes with a bit drop of 1 m or more, loaders are used in combination with manual sand feeding. For boundary boreholes with a bit drop of less than 1 m, small flow grouting is used in combination with manual sand feeding. The sand ratio of boundary boreholes is controlled at more than 50%. The increase in the proportion of wind-blown sand is conducive to increasing the concentration of the mixed filling material and reducing its flow performance, thereby reducing the loss of the filling material expanding out of the governance boundary.
- Adopt intermittent grouting and sand feeding technology. In the process of grouting and sand feeding of the boundary boreholes in the goaf area, intermittent procedure is adopted. In the intermittent process, the filling material injected into the goaf has reduced flow performance under the action of physical deposition and chemical consolidation, and finally accumulates in the area close to the discharge port of the goaf borehole. Generally, for drilling holes at the boundary of goaf, the intermission time is 2 h after the single grouting volume does not exceed 50 m^3. For drilling boreholes with partial filling volume exceeding 300 m^3 and grouting not yet terminated, the intermission time and intermission frequency are increased. The single grouting volume does not exceed 30 m^3, and the intermission time is not less than 4 h.
- Add quick-setting agent to curtain borehole grouting. During the curtain borehole grouting process, evenly add 2% of cement weight of quick-setting agent in the secondary mixing tank. For drilling boreholes with partial filling volume exceeding 300 m^3 and grouting not yet terminated, appropriately increase the proportion of quick-setting agent to 3% of cement weight.

Grouting of Grout Boreholes

Before each grouting operation, inject 2–3 m^3 of clean water to clean the rock cracks and facilitate the dispersion of slurry. Each grouting should be conducted in a thinner (low concentration) and thicker (high concentration) manner. After the grouting starts, it is recommended that the following actions be taken:

- Observe the slurry suction volume and pump pressure of the pump
- Record various phenomena occurring during the grouting process
- Collect original data
- Follow the direction of the lead engineer according to the actual grouting situation and adjust the grouting volume and slurry concentration in time.

When the grouting volume is large, intermittent grouting method should be used for construction, or 2% of cement weight of accelerator should be added to the slurry. The grout and sand injection process in the goaf is generally divided into three stages according to the situation of the drilled hole: air-suction stage, non-air suction stage, and slurry return stage. In the first stage, the slurry flows into the hole at high speed, resulting in negative pressure at the borehole mouth. A forklift is used to manually add a large amount of sand. In the non-air suction stage, the space in the hole becomes smaller with the grouting operations. The slurry flow rate in the borehole slows down, and the borehole mouth does not absorb air. At this stage, manually feeding sand is used. In the last stage when the air pressure in the borehole is greater than the downward pressure of the slurry, the borehole mouth appears to return slurry. At this stage, the equipment is replaced for pressure grouting.

The sand should be added in an open process. The sand added by the sand thrower is brought into the hole by the flow of slurry. To prevent the hole from being blocked, a large amount of sand should be avoided in a short time. The amount of sand added should be controlled at 20% of the total grouting filling volume, which can be adjusted according to the actual situation on site. If the drilling borehole is blocked during the air suction stage or the non-air suction stage, the borehole needs to be swept.

Grouting is terminated when the unit grouting volume is less than 50 L/min and is stable for more than 15 min under the end pressure specified in the design. Before, after and during grouting, the grouting pump pressure, orifice pressure, grouting volume, slurry concentration and adjacent hole water level should be regularly observed and recorded.

After the grouting project is completed, the operation data should be compiled in a timely manner. The completion data should include drilling records, grouting records, inspection and testing data of raw materials and slurry. Figure 3.3 shows the processes of the grouting operation.

Handling of Special Situations

Unexpected situations are often encountered in execution of field operations. The drilling operations cannot be conducted to reach the designed depth if borehole collapses occur, drill rods get stuck, or serious leakage occurs. One of the solutions to these problems is to adopt the top-down segmented grouting process. The process

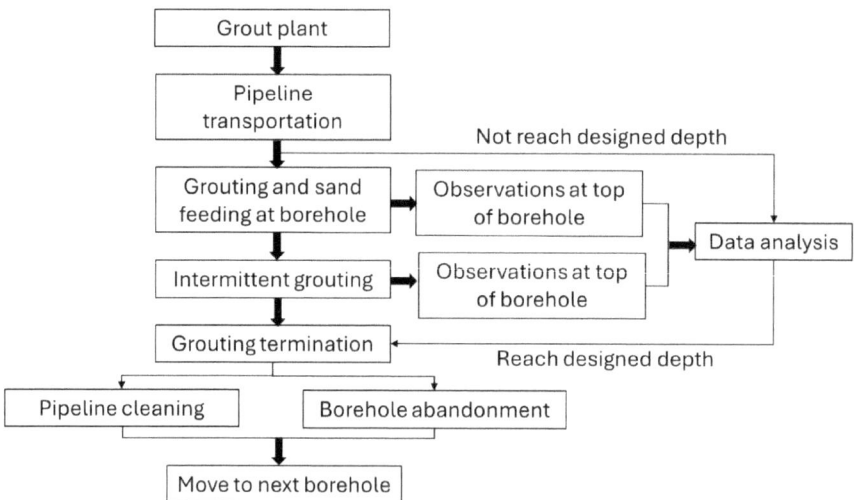

Fig. 3.3 Flowchart of grouting operation

includes alternating grouting and borehole cleaning until the designed borehole depth is reached, and finally grouting of the goaf is carried out.

Grouting operation is often challenged with daylighting of grout. Low pressure, flow limiting, and limited quantity and intermittent injection can generally be used to deal with it. If necessary, caulking and surface sealing force methods should be used.

When grouting in one borehole moves through other boreholes, grouting should be conducted simultaneously with one pump in one borehole if the grouting borehole meets the grouting conditions. Otherwise, the boreholes that inadvertently receive grout from another injection borehole should be blocked. After the grouting of the grout hole is completed, the blocked boreholes should then be swept and flushed to the designed depth. Then grouting operations can be conducted.

Grouting should be conducted continuously. If it is interrupted for some reason, grouting should be resumed as soon as possible. Otherwise, the borehole should be flushed immediately. Then resume grouting. If flushing is not possible, the hole should be swept, and grouting should be resumed.

If the grouting volume of a single hole is exceeded and the grouting is difficult to end normally, low-pressure, flow-limited, quantity-limited, and intermittent grouting methods can be used to grout. After reaching the designed grouting termination standard, the grouting of this section can end.

3.3.2.5 Remediation Sequence

The remediation sequence should be conducted in the order of curtain first, grouting in goafs later and edge first, and interior later. Drilling is conducted after the area is divided for the purpose of sequences and intervals. Each area is conducted in accordance with the principle of periphery first and interior later. It is anticipated that the grout in the first sequence boreholes spreads over a larger area. The second sequence grout boreholes will fill the remaining voids left from the first sequence. Such a sequence maximizes the grout filling rate.

If there is grouting interference between grout boreholes in the same sequence, the construction of the corresponding subsequent holes can be cancelled. The drilling operation of the same sequence should be conducted in accordance with the principle of first low bedding angle and then high bedding angle. For boreholes of different sequences, the drilling operation of the subsequent sequence boreholes is conducted only after all the boreholes in the previous sequence are completed.

3.3.3   Monitoring

3.3.3.1   Deformation Monitoring

The monitoring points are arranged in the 4^{-3} and 5^{-2} coal seam goaf areas. Two monitoring sections are arranged respectively. The spacing between the monitoring points is approximately 20 m. Sixteen monitoring points are installed. The locations of the monitoring points are shown in Fig. 3.4.

The monitoring point markers are prefabricated reinforced concrete with a cross in the center of the top of the markers. The markers are established according to the specifications of the fourth-class leveling point markers in the Engineering Surveying Code (GB50026-2007 with clause explanations). The buried depth of the markers is greater than 0.25 m, which is the local frozen soil depth.

The deformation monitoring parameters in the goaf area include vertical displacement and horizontal displacement. Regular visits are made at each measuring point. The elevation and plane coordinates of the measuring points are measured to calculate the vertical displacement and horizontal displacement of the surface of the project area.

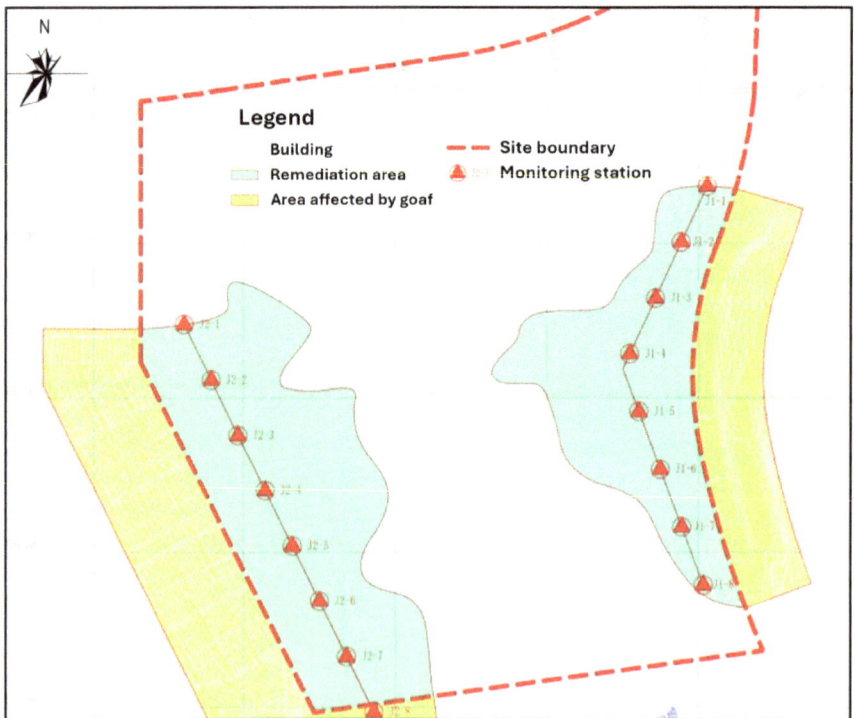

Fig. 3.4 Site layout and locations of monitoring points

The monitoring starts one month before the goaf area remediation and continues throughout the entire remediation, and one-year remedial action—operation after remedial action—construction. The monitoring frequency is twice a month before remedial action, once a day during the remediation action—construction period, and once a month during remedial action—operation period. The entire monitoring period lasts approximately two years, and 576 data points are collected.

Monitoring accuracy is specified in the work plan. According to Article 10.1.3 of the Engineering Surveying Standards (GB50026-2007 with clause explanations), the monitoring level of the roadbed section is Grade IV, and the monitoring level of the bridge section is Grade II.

3.3.3.2 Water Quality Monitoring

Groundwater samples are collected for water quality monitoring. The samples are measured for color, pH, smell, turbidity, visibility. One or two groups of water samples are collected before remedial action, one group of water samples are collected every 2 weeks during remedial action—construction, and 1 group of water samples is collected every month after construction. A total of 90 groups of water samples are collected.

3.3.4 Post-construction Quality Inspection

3.3.4.1 Inspection Parameters

The post-construction quality inspection is conducted by a third-party engineering entity. Based on the geological structural characteristics of the goaf area, engineering geological conditions and previous work experience, combined with the local actual situation and project progress requirements, the quality inspection scope includes six aspects: construction technical data inspection, drilling inspection, grouting inspection, grout consolidation strength inspection, wave velocity test, and borehole televiewing inspection.

- Construction technical data inspection: Inspect the technical data generated during the construction process of the project. The data should be timely, detailed, objective, and can truly reflect the construction process. There should be no false or supplementary phenomena. Arrange the inspection drilling based on the data inspection and analysis.
- Drilling inspection: Drilling and coring testing is carried out within the treated goaf area. The hole is opened at 150 mm, and the diameter is changed to 91 mm at 5 m below the bedrock. The core recovery rate of each bedrock section should be greater than 90%. The slurry stone body in the goaf and its caving section should be noticeable. There should be no drill drop, no leakage of circulating fluid, and

no blowing and suction during the drilling process. During the construction of the test hole, various conditions such as the core recovery rate, circulating fluid consumption, and footage speed should be recorded in detail, and a bar chart of the test hole should be drawn.

- Grout consolidation strength inspection: The slurry stone body obtained by drilling and coring should be sent to the laboratory for indoor unconfined compressive strength test in accordance with the Testing Procedures for Cement and Cement Concrete for Highway Engineering (JTGE30-2018) after standard maintenance for 72 h. The compressive strength of the stone body is required to be $\geq$ 2.0 MPa.
- Wave velocity test: Use ultrasonic or high-power sonic instrument to conduct comprehensive logging before and after grouting. According to the Code for Seismic Design of Buildings, the average shear wave (transverse wave) velocity of the grouting treatment section in the borehole V_{sm} is required to be $\geq$ 350 m/s.
- Borehole televiewing inspection: Describe the borehole in detail and intuitively. The formation fissures in the goaf and its caving section should be filled with slurry and stone bodies, and there should be no cracks or cavities.
- Grouting test: Conduct grouting test on the inspection boreholes, record the grouting volume and judge the filling density according to the grouting volume.

3.3.4.2 Quality Inspection Point Layout

The inspection holes should be evenly distributed in the project area. At the same time, they should be selected according to the importance of the project in important locations with high allowable deformation requirements such as important buildings (structures); locations with complex geological conditions such as rock mass fragmentation, hole collapse, and drill drop; near the hole section with large injection volume of the last sequence hole; and locations where the grouting situation is abnormal and where analysis shows that there may be problems with the grouting quality.

According to the requirements of GB 51180-2016, the number of inspection holes is 5% of the number of grouting holes. In this project, 13 inspection boreholes are selected.

3.3.5 Remedial Action Construction Organization

3.3.5.1 Project Management Approach

Figure 3.5 shows the project team and required disciplines.

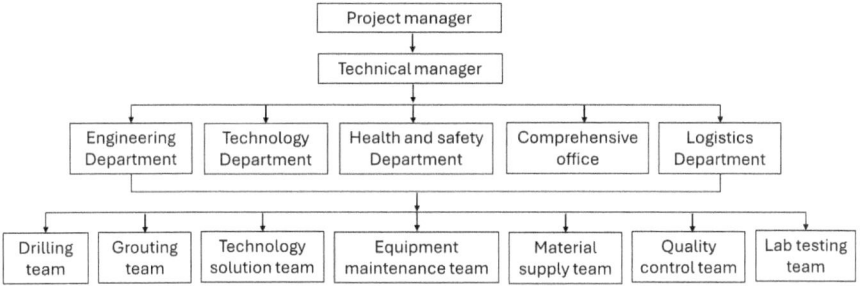

Fig. 3.5 Project management approach

3.3.5.2 Main Construction Equipment

The main mechanical equipment used in the construction is summarized in Table 3.1.

3.3.6 Remedial Action—Construction Duration

This project started on June 18, 2020. On September 27, 2020, temporary construction, drilling, grouting, sand filling, water quality monitoring, and deformation monitoring were completed in the treatment area. The actual construction period is approximately 100 days. According to the requirements of the design documents, the remedial action—operation start after 3 months for quality inspection. The inspection started on December 21, 2020, and the inspection and re-grouting work were completed on January 15, 2021.

3.3.7 Variations from Engineering Design

During the drilling process, the drilling team found that the average drill drop height and recovery rate of the goaf in the treatment area were greater than the corresponding parameters in the design documents, resulting in an increase in the volume of the voids in the treatment area. A change in the grouting and sand filling volume was proposed and submitted in a project variation application form. The engineering department reviewed the variation application form and issued a design change notice to change the grouting and sand filling volume with updated drawing design and related budget documents.

Table 3.1 Main construction equipment invested in this project

Classification	Machine name	Specifications	Quantity
Temporary construction equipment	Transformer	S11-M	1
	Primary distribution box	1700 × 700 × 350	1
	Secondary distribution box	500 × 600 × 180	3
Drilling engineering equipment	Geological drilling rig	100T	8
	Electric welding machine	XY-200, truck-mounted drilling rig	2
	Cutting machine	400 type	1
Grouting engineering equipment	Ash (mud) mixer	LK-2	6
	Fly ash and cement tank	100T	3
	Swirl conveyor	LS-200-1800	3
	Feeding control system	2T	3
	Grouting flow meter	DN32	5
	Grouting pump	BW-450/250	5
	Water pump	75WQ50-10-3	4
Test equipment	RTK	SET2000	1
	Level	DS03	1
	Inclinometer	JTL-50	1
	Viscometer	1006	1
	Hydrometer	National standard	1
	Rack balance	HC-TP-12	1
	Electronic balance	MP2002	1
	Digital drying oven	101A-1	1
	Bench scale	25 kg	1
	Vicat instrument	CHN-1	1
	Maintenance pool	4 × 1.2 × 0.3	1
	Trial mold	7.07 × 7.07 × 7.07	12
	Aluminum box		40
	3D laser scanner	C-ALS	1

3.4   Quality Control and Quality Evaluation

3.4.1   Quality Control of Key Operations

3.4.1.1   Site Preparation Works

During the construction preparation stage, our unit was equipped with temporary works that matched the on-site engineering volume and construction schedule, mainly including project department construction, slurry station construction, material storage yard construction, water reservoir construction, road construction, etc. Comprehensive construction facilities in the treatment area. The comprehensive construction facilities integrate slurry making, grouting, material and equipment storage, pipe processing and construction management, including one large grouting station (slurry making and grouting capacity reaches 1500 m³/day), one 100-ton cement storage tank, 100 m² warehouse for cement in bags, two 100-ton fly ash storage tanks, 400 m³ water tank, large steel stacking and processing workshop, construction equipment maintenance workshop, laboratory, construction unit and supervision unit on-site office and other facilities.

The number of temporary project offices and conference rooms meets the needs of on-site construction management personnel. The storage tanks, storage yards, and water reservoirs match the pulping capacity of the pulping station. The pulping station capacity can ensure that the overall construction project is completed according to the designed construction period.

3.4.1.2   Drilling Operation

- Use RTK field measurement and layout. The borehole position deviation is ≤ 2 m. Accurately and timely record the construction measurement data.
- Final borehole depth and final borehole layer: Drill to the target coal seam floor or the residual cavity floor of the goaf 1.0 m below the final hole; or when the expected depth is reached but the final hole requirements are still not met, communicate with the design unit to determine whether to continue drilling.
- Final borehole inclination: The borehole inclination does not exceed 1.5° every 100 m.
- During the drilling process, mud wall protection is used in the Quaternary loose layer. Clear water drilling is used in the bedrock. Record drilling progress and catalogue cores. During the drilling process, the depth, layer and flushing fluid consumption are recorded in detail if water leakage, drill drop, or buried drill occurred.
- All treatment boreholes are cored in the whole hole, and the coring rate meets the requirements of the design documents.
- The orifice pipe is cast using the "sitting grouting method". The orifice pipe is a φ133 mm steel pipe that is 5–6 m below the stable bedrock. The cement slurry

water–solid ratio is 1:1.5–1:2, and 3% of the cement weight of water glass is added.

- After the orifice pipe is cast, it is accepted by the on-site technicians and the grouting strength reaches the design strength requirement before grouting can begin.
- After reaching the drilling final hole standard, the person in charge of the drilling project of the project department, together with the professional supervision engineer and the representative of the owner's engineering department, will accept the various indicators of the drilling, check the on-site construction data, measure the drilling rig, and check the core. After the on-site acceptance and the "drilling construction data" acceptance are all qualified, the single hole acceptance can be included in the project volume. During the acceptance work, all holes passed the three-party acceptance and met the qualified standards.

The parameters and indicators of each hole in the drilling project meet the requirements of the construction drawing design of goaf control and relevant specifications. The parameters of each hole are shown in the application form for single hole drilling inspection. A total of 252 copies of the application form for single hole drilling inspection and related attachments have been submitted for review of the control project, including the drilling and casting of the boreholes and orifice pipes of 252 control boreholes.

3.4.1.3   Grout-Making Operation

Grout-Making Quality Control

- The slurry mix ratio of water: solid phase is 1:1–1:3, and the cement: fly ash mix ratio is 2:8–3:7. The final mix ratio shall be subject to the mix ratio report issued by the laboratory that meets the design requirements.
- A laboratory shall be built at the construction site, equipped with a standard curing box, a hydrometer, an electronic scale, a measuring cylinder, a beaker and other test equipment. Before and during construction, the test personnel shall conduct slurry performance measurement tests and slurry mix ratio tests under the joint witness of our quality inspection personnel, supervision personnel and representatives of the construction party. The test contents include slurry mix ratio, specific gravity, initial and final setting time, stone rate and stone body strength.
- The slurry mix ratio shall be strictly followed, and various indicators of the slurry shall be randomly checked.
- Raw materials: Water is measured by container. Cement is measured by bag or weighing system. Fly ash is measured by a weighing system.
- The slurry process shall be strictly conducted according to the slurry process flow chart.

- Mixing process: The mixing time of each slurry in the first-level mixing tank shall not be less than 3 min to fully mix the water, cement and fly ash. The second-level mixing tank shall continue to mix to prevent the slurry from settling.
- Accurately and timely record the slurry making data.

Evaluation of the Quality of the Slurry Making Operation

The cement and fly ash materials used in the slurry making were sampled and evaluated under the witness of a professional supervision engineer. The test results all met the requirements of the "Construction Drawing Design of Goaf Management" and relevant specifications (see the "Drilling Grouting Inspection Application Form Attachment" and "Material Inspection Application Form Attachment" in the "Completion Data" for various parameters). During the slurry making construction, the project department's slurry making project manager, professional supervision engineer, and owner's representative "three parties 24 h a day in two shifts" controlled every link of the slurry making. The loading amount, mixing time, and mixing standard operation of each slurry plate have all passed the acceptance of the three parties. Each slurry plate can only be included in the project volume after it meets the qualified standards.

3.4.1.4 Grouting Operation

Grouting Operation Quality Control

- Grouting follows the principle of "washing the hole first and then grouting". Before grouting, clean water is used to wash the hole for about 5–10 min, and then the slurry with the specified ratio is injected. For grouting holes with blockage and sudden pressure at the hole mouth, the grouting holes are repeatedly swept, penetrated, diluted or washed with high pressure water, and grouting is repeated to ensure that each hole is truly filled with grouting. For large holes with lost drilling, intermittent and repeated grouting is used for grouting.
- When grouting, avoid injecting a large amount of cement fly ash slurry in a short period of time. When the grouting volume is large, intermittent grouting is used for construction. During the construction process, the design unit, supervision unit and construction unit organized on-site meetings in a timely manner according to the problem of large slurry volume in some drilling holes during the construction process and formulated corresponding intermittent grouting plans according to different drilling types to ensure the quality and efficiency of grouting.
- At the end of the grouting of the grouting hole, the pump pressure gradually increases. When the grouting pressure exceeds 1 MPa, the pump volume is less than 50 L/min and is stable for more than 5 min, the grouting of the hole is completed. If a large amount of grouting occurs in the surface cracks during the

grouting process, the intermittent grouting process should be adopted immediately. When the grouting reaches the design end standard, the grouting of the hole is completed.

- Record the grouting construction data. The record sheet should truthfully reflect the grouting time, material ratio, slurry specific gravity, consistency, material consumption, pump pressure, pump volume, orifice pressure, grouting volume, abnormal phenomena during the grouting process, treatment measures and their effects.

- In the grouting project, the person in charge of the grouting project of the project department, together with the professional supervision engineer and the representative of the owner's engineering department, the "three parties" monitor the grouting situation of the grouting point in real time. When the conditions for terminating grouting are met, the "three parties" accept the various indicators of the grouting point and check the construction data. After the on-site acceptance and the "grouting construction data" are accepted, the grouting volume of this grouting point can be included in the "grouting metering table". During the acceptance work, all grouting points passed the acceptance of the "three parties" and reached the qualified standards.

Grouting Operation Quality Evaluation

The parameters and indicators of each hole of the grouting project meet the requirements of the construction drawing design of goaf treatment and relevant specifications. The parameters of each grouting point can be found in the drilling grouting inspection application form. A total of 250 copies of the drilling grouting inspection application form have been submitted for review for the treatment project, including grouting records and grouting results of 250 treatment boreholes.

3.4.2 Quality Control of Construction Materials

3.4.2.1 Material Supply

Grouting materials mainly include water, cement, fly ash, and water glass. The construction water of the project is taken from the well near the site and transported to the construction site by pipeline. The cement used in the construction of the project is produced by a local cement plant. The fly ash used in the construction of the project is produced by a local Power Plant. The water glass used in the construction of the project is a product purchased from the market that meets the design requirements.

3.4.2.2   Material Inspection

Before the construction of the goaf control project, we inspected and tested the water source, and the water quality meets the requirements for construction water. The water used in the construction is transported to the project water reservoir by Φ108 steel pipe and is not polluted. Cement sampling was conducted 20 times, and the samples were sent to the city engineering quality inspection station for testing. The test results all met the requirements of the goaf area remediation design, relevant specifications, and quality control of this project. The fly ash sampling was conducted 42 times, and the samples were sent to the city engineering quality inspection station for testing. The test results all met the requirements of goaf area remediation design, relevant specifications, and quality control of this project. The water glass sampling was conducted 2 times, and the samples were sent to a certified laboratory for testing. The test results all met the requirements of goaf area remediation design, relevant specifications, and quality control of this project. The slurry ratio inspection was conducted 12 times, and the samples were sent to a certified laboratory for testing for testing. The test results all meet the requirements of the goaf area remediation design, relevant specifications, and quality control of this project. The water quality monitoring was inspected 11 times, and the samples were sent to the Shaanxi Provincial Drinking Water Supervision and Inspection Station for testing. The test results all meet the requirements of the goaf area remediation design, relevant specifications, and quality control of this project.

3.4.2.3   Quality Inspection of Intermediate Products

- Grout performance index: The project department quality inspection personnel, together with professional supervision engineers and representatives of the owner, witnessed the grout performance index test. The on-site grout specific gravity and stone rate were tested more than 200 times, and all passed.
- Grout test block compressive strength: The project department quality inspection personnel, together with professional supervision engineers and representatives of the owner, witnessed the production of more than 200 sets of 70.7 mm by 70.7 mm by 70.7 mm test samples. Standard maintenance was conducted and 106 sets of test samples were sent for inspection to measure the stone compressive strength at 28th day. The test results all met the requirements of goaf control remediation design, relevant specifications, and quality control of this project.

3.4.3   Conclusion of Post-construction Quality Inspection

The inspection work completed 13 measurement points, drilling and borehole tele-viewing, and wave velocity testing in 13 boreholes with a linear footage of 539.6 m. Also completed are nine groups of uniaxial compressive strength tests and one construction data inspection and evaluation. The results of drilling and borehole tele-viewing show that all inspection boreholes have no obvious drill drop phenomenon (JC1-2 drill drop 0.2 m, but grouting test injection 8.6 m^3), the underground coal seam goaf of the site has been effectively filled. Results from the televiewing and coring of the consolidated slurry indicate that the grout is in full contact with the goaf roof.

The wave velocity test results show that the wave velocity of all test holes in the grouting section is > 1000 m/s, which meets the design requirements. The strength test of the grouting stone body shows that the mechanical strength index of the stone body meets the design requirements. The grouting test shows that the average grouting volume of the 13 test holes is less than 5% of the average grouting volume of a single hole during the construction period, and the overall treatment effect is good. The JC1-2 hole with construction quality defects was evaluated by grouting, and the cumulative grouting volume was 8.6 m^3. The goaf was further reinforced to ensure the safety of the site. This grouting project effectively filled the underground void area and achieved good treatment results. The project quality assessment score was 96.88 points, indicating that the project quality was qualified.

3.5 Progress Management of Construction Process

Before entering the site, the project team ensures the construction period by formu-lating a construction progress plan, a main labor ratio plan, a main machinery and equipment, instrument and construction machinery entry plan, and a main mate-rial supply plan. During the construction process, our team implements dynamic progress management measures based on the actual construction situation to ensure construction quality and progress. The main management measures are as follows:

- Hold on-site progress meetings regularly to strengthen the management of the progress of the construction links. Through early prediction, node coordination, and timely analysis of drilling, construction, materials and external environment, timely adjustments are made to possible construction period risks, and at the same time, the pre-control of construction period costs is strengthened.
- Implement dynamic tracking and adjustment of key links to timely pre-control progress risks.
- In order to ensure that the construction tasks are completed on schedule, the project department promptly reports the monthly progress and progress payment approval, focusing on giving priority to funding and timely payment of outsourced construction workers.
- For drilling and construction projects with high coordination difficulty, we scien-tifically arranged the number of drilling rigs, spent time and effort on early planning, added grouting systems, and provided technical support for the progress.

Fig. 3.6 Project safety and health approach

3.5.1 Safety Management During Construction Process

3.5.1.1 Safety Assurance System

The project manager is the first person in charge of safety, the deputy project manager is the manager of the safety of the entire project, the sub-project manager is the person in charge of the safety of each sub-project, and each shift has a safety manager who is the executor of safety; the safety assurance system is shown in Fig. 3.6.

3.5.1.2 Safety Assurance Measures

The project department shall establish a safety management leadership organization and establish and improve the safety production responsibility system of departments at all levels, with responsibilities assigned to individuals. The construction site and construction site shall implement fully enclosed safety management to prevent entry and interference by unauthorized personnel. For warehouses or storage points for materials and equipment, and living areas for construction personnel, dedicated personnel shall be on duty to prevent fire and theft. Electricity for construction. The cables in the distribution box shall be provided with casings, and the wires shall not be in disorder. Outdoor branches shall be suspended and shall not be dragged or tied to scaffolding. Outdoor lighting fixtures shall not be less than 6 m from the ground, and indoor lighting fixtures shall not be less than 2–4 m from the ground. High-voltage wires, and transformers at and around the construction site shall have eye-catching safety warning signs. Especially during nighttime construction, specific measures shall be taken to prevent electric shock accidents.

It is essential to establish a regular safety inspection system. There shall be time and requirements, and key areas and dangerous positions shall be clearly identified. Safety inspections shall be recorded, and hidden dangers found shall be rectified in a timely manner, with designated personnel, time and measures. The project manager is responsible for production safety and safety education. A production safety meeting is held every week to check the implementation of production safety regulations and establish a production safety reward and punishment system so that everyone pays attention to safety. During the construction, our project department strictly implemented various production safety technical standards, effectively grasped the key points that are likely to cause safety problems during construction and prevented the occurrence of production accidents. As of the end of construction, no personal injury or death accidents occurred in this project, and the mechanical equipment accident rate was controlled within 5‰.

3.6 Summary

Based on the test results, the overall remedial action—construction quality of this project is achieved the purpose of goaf treatment, proving that the construction technology of this project is successful and effective. The treatment area of this project is relatively large, and the geological conditions are complex. There are two layers of coal mined in the area. In addition, there are small coal mines that use room-and-pillar mining methods and large mines that use mechanized comprehensive longwall mining methods. It is difficult to fully and accurately control the coal seam depth and goaf conditions of the entire treatment area with limited exploration drilling holes. Therefore, the estimated drilling and grouting engineering volume is inevitably different from the actual situation.

The surface subsidence monitoring is recommended to be strengthened around critical buildings. It is also critical to perform the deformation monitoring of deep bedrock in the goaf area. If abnormal measurements are detected, a systematic analysis should be immediately conducted on the underground goaf, backfill foundation, and infrastructure. Appropriate remedial measures should be taken to mitigate any risk of ground subsidence.

Chapter 4
Application of Paste Materials in Mitigation of Subsidence Caused by Room-and-Pillar Mining

Abstract The surface deformations caused by shallow goafs can result in serious damage to buildings (structures) during mining. The movement duration is typically short, and the underground void rate and residual deformation are relatively small. The collapse pits formed on the surface of shallow unconventional room-and-pillar goaf can cause serious damage to ground buildings. The underground voids and residual deformation left in the goafs may also pose a potential hazard to surface stability. A paste filling technology is successfully used to remediate the shallow goafs. Paste filling materials have the characteristics of high solidity, high strength and good economic benefits. They have great advantages for goafs in coal mines where the roof rock strata are relatively complete and hard, the goaf space is relatively large, and the roof has not collapsed or has not completely collapsed.

Keywords Paste filling technology · Room-and-pillar mining · Underground voids · Ground deformation

Coal resources are rich in northwest China. The geological conditions of the coal resources are also favorable to mining such as shallow burial depths and thick coal seams. According to the national coalfield survey, Xinjiang, Inner Mongolia, Shanxi, and Shaanxi provinces account for approximately 81% of the total national resources. In the past decade, these regions have integrated and established many advanced large-scale modern mines, which have maximized the exploitation of coal resources and increased economic benefits. However, in the process of integration and expansion of large coal mines, mined-out voids or goafs that are associated with previous small-scale coal mines need to be properly managed. According to the ratio of mining depth to mining height, goafs can be divided into shallow goafs, medium-deep goafs, and deep goafs.

The mining depth to height ratio is less than 40 for the shallow goafs. The shallow goafs that are caused by longwall mining may result in dynamic surface movement, large vibration speed and large deformation. Step-shaped collapses, earth fissures, or collapse pits may appear on the ground. The deformations on surface can cause serious damage to ground buildings (structures) during mining. The surface fissures

S. Dong et al., *Prevention and Reclamation of Mining-Induced Land Subsidence*, SpringerBriefs in Earth Sciences, https://doi.org/10.1007/978-3-031-78158-2_4

may be connected to the lower fracture zone. Fortunately, the movement duration is short, and the underground void rate and residual deformation are relatively small. The collapse pits formed on the surface of shallow unconventional room-and-pillar goaf can cause serious damage to ground buildings. The underground voids and residual deformation left in the goafs may also pose a potential hazard to surface stability. The type of shallow goafs accounts for approximately 25% of coal mine goafs in China.

The mining depth to height ratio is greater than 40 but less than 200 for the medium-deep goafs. The surface of the medium-deep longwall mining collapse goaf may produce different degrees of movement, deformation and cracks. These deformations may cause different degrees of damage to ground buildings (structures). The fractures induced by longwall mining and cavities left from room-and-pillar mining may cause deformations in the overburden, which may pose different degrees of potential hazards to the stability of the surface. The type of medium-deep goafs accounts for approximately 60% of the goafs in China's coal mines.

The mining depth to height ratio is equal to or greater than 200 for the deep goafs. The surface movement range of deep goaf is relatively large, but the movement speed is slow. The movement deformation is relatively small. If there is no special adverse terrain and geological structure, there are generally no obvious collapse cracks on the surface, even for the longwall mining method. During the mining process, structural damage to buildings is typically not obvious. The remaining cavities left from room-and-pillar mining or fractured zones induced by longwall mining generally do not pose a potential hazard to the stability of the surface. The type of deep goafs accounts for approximately 15% of the goaf in China's coal mines. The proportion of deep goafs tend to increase in the future.

4.1 Characteristics of Room-and-Pillar Goaf in Coal Mines in Northern Shaanxi

4.1.1 Regional Geological Conditions

The Jurassic coalfield where the Yushenfu mining area is located on the eastern wing of the Ordos syncline and the northern Shaanxi slope. According to regional data, the basement of the northern Shaanxi slope mainly contains Wubao-Jingbian EW structural belt, Baode-Wuqi NE structural belt, and Yulin West-Shenmu NE structural belt. These structural belts have a certain control on the formation and distribution of coalfields.

The Yushen mining area is in the eastern part of the Jurassic coalfield in northern Shaanxi. The strata are generally oriented in NWW direction. The monocline structure is inclined to the north, and the dip angle is less than 1°. Only some parts of the mining area have gentle undulations of varying sizes, but no large faults with a drop of more than 20 m are identified. There is no magmatic activity. The tectonic movements in the mining area are mainly vertical movements, forming a series of parallel unconformities. The unconformities include the parallel unconformities between the following formations:

- The Yongping Formation (T3y) of the Upper Triassic and the Fuxian Formation (J1f) of the Lower Jurassic
- The Yan'an Formation (J2y) of the Middle Jurassic and the Fuxian Formation (J1f) of the Lower Jurassic
- The Yan'an Formation (J2y) and the Zhiluo Formation (J2z)
- The Zhiluo Formation (J2z) and the Anding Formation (J2a)
- The Anding Formation (J2a) and the Luohe Formation (K1L) of the Lower Cretaceous.

The stratigraphy of the mining area belongs to the Ordos Basin division of the north China stratigraphic area. Based on geological mapping and drilling, the strata in the mining area are as follows:

- Yongping Formation of the Upper Triassic (T3y)
- Fuxian Formation of the Lower Jurassic (J1f)
- Yan'an Formation of the Middle Jurassic (J2y)
- Fuxian Group of Lower Jurassic (J2f)
- Zhiluo Formation (J2z)
- Anding Formation (J2a)
- Neogene Upper Pliocene Baode Formation (N2b)
- Quaternary Lower Pleistocene Sanmen Formation (Q1s)
- Middle Pleistocene Lishi Formation (Q2l)
- Upper Pleistocene Salawusu Formation (Q3s)
- Malan Formation (Q3m)
- Holocene aeolian sand (Q4eol)
- Alluvial deposits (Q4al).

Table 4.1 provides the details of the strata in the mining area.

4.1.2 Mining Characteristics

Coal in this area is extracted by the room-and-pillar mining method. In the room-and-pillar method, the coal is mined in rooms of 5–9 m wide separated by pillars which may or may not be partially extracted later. Initially, a series of parallel entries are driven through the seam with interconnecting openings (breakthroughs) driven at right angles through the pillars between the rooms. Such a checkerboard pattern of openings is advanced through the coal seam to the limit of the area planned for mining. At this point approximately 50% of the coal will have been mined. The coal pillars between adjacent rooms may be fully or partially removed (extracted) during

Table 4.1 Strata in the study area

Group	Lithological characteristics	Thickness (m)	Distribution
Aeolian sand (Q4eol), Alluvial deposits (Q4al)	• It is dominated by modern aeolian sand, mainly medium-fine sand and sub-sandy soil • There are alluvial and diluvial layers in river valley beaches and some low-lying areas	Aeolian sand 0 ~ 28.00 5.00 Alluvial deposits 0 ~ 22.00 12.00	Aeolian sand is distributed on Liangmao, and alluvial layer is distributed in the valley
Upper Pleistocene (Q3), Malan Group (Q3m), Sarawusu Group (Q3s)	• Grayish yellow to grayish brown sub-sand and silt, homogeneous, loose, and with large porosity • Grayish yellow to brownish black fine sand, sub-sand soil, sandy clay, with gravel at the bottom	Malan Group 1 ~ 15.00 10.00 Sarawusu Group 0 ~ 65.00 20.00	Distributed in Shigetai and south of Ningtiaota
Middle Pleistocene (Q2), Lishi Group (Q2l)	Light brown to yellowish brown sub-clay and sub-sandy soil, interbedded with silty sand layers, paleosol layers, calcareous nodule layers, and a gravel layer at the bottom	0 ~ 78.00 42.00	Widely distributed in Liangmao hilly areas
Lower Pleistocene (Q1), Sanmen Group (Q1s)	Brownish red to light flesh-red sub-clay and gravel layers, intercalated with calcareous nodule layers	0 ~ 50.00	Distributed in north of Daliuta
Upper Pliocene (N2), Baode Group (N2b)	Brownish red to purple clay or sandy clay, intercalated with calcareous nodule layers, containing vertebrate fossils	0 ~ 110	Distributed in Miaogou and Zhugaigou areas
Middle Jurassic (J2), Anding Group (J2a), Zhiluo Group (J2z)	The upper part is mainly composed of purple-red to dark purple mudstone and sandy mudstone, while the lower part is mainly composed of purple-red medium-coarse-grained feldspar sandstone Purple mixed ~ mudstone, sandy mudstone, sandstone, and sometimes conglomerate at the bottom	Anding Group 0 ~ 98.66 47.63 Zhiluo Group 0 ~ 137.54 92.98	Distributed in Shenmu Wotucaidang Distributed in Shenmu Yao Town
Middle Jurassic (J2), Yan'an Group (J2y)	Light gray to dark gray sandstone, mudstone, and sandy mudstone contains multiple mineable coal seams. They are the main coal-bearing strata in the basin, with a maximum of 13 layers, usually 3–6 layer, the maximum total thickness that can be mined is 27 m, the maximum thickness of a single layer	0 ~ 245.93 205.61	Distribution across the entire region

(continued)

Table 4.1   (continued)

Group	Lithological characteristics	Thickness (m)	Distribution
Lower Jurassic (J1), Fuxian Group (J2f)	Purple-red, gray-purple and gray-green sandy mudstone, interbedded with black mudstone, thin coal line, oil shale and quartz sandstone, and the bottom is fine to coarse conglomerate	$\frac{0 \sim 37.17}{9.73}$	Distributed intermittently in the area
Upper Triassic (T3), Yongping Group (T3y)	Grayish white to grayish green thick layered fine-grained feldspar quartz sandstone, interbedded with grayish black to blue-gray mudstone and sandy mudstone, containing thin coal lines, and is an oil-bearing stratum	80 ~ 200	Exposed in the areas of Kuye River and Tuwei River

final, or retreat, mining. After full pillar removal, the rock above the mine collapses and the overburden gradually settles, creating surface fissures and subsidence.

According to the survey, most of the small-scale coal mines in the area adopt room-and-pillar mining. However, the room-and-pillar mining method is currently an unconventional coal mining method used by small coal mines. It is generally used in the goafs of ancient kilns in which a tunnel network of 20–30 m wide is established. The tunnels are divided into approximate squares, and then mined manually. Coal pillars are left at the four corners of the mining room to support the roof, and the roof falls naturally after the roof is released. This method of coal mining has low equipment investment and a short construction period, but the resource recovery rate is low. According to investigations, the recovery rate of small coal mines currently used in northern Shaanxi with room-and-pillar coal mining is below 30%, resulting in a huge waste of resources. At the same time, it also leaves serious hidden dangers for subsequent ground construction and underground production.

The room-and-pillar mining operations are not standardized in China. The recovery rate is low, and the regularity of surface movement and deformation is poor. Generally, shallow goaf areas are prone to collapse pits or annular settlement areas, while deep goaf areas have slow surface movement and deformation, and the residual void rate is large. In general, the room-and-pillar methods used in the Yushenfu mining area include the following.

Conventional room-and-pillar mining method: The Yongle Coal Mine in Yuyang District, located in a sand-based submerged coal mining area, adopts the room-and-pillar mining method. Two 150-m long tunnels are dug on one side of the main tunnel. The drift (2.2 m high, 4 m wide) forms a 40 m in the middle of the section, along the advancing direction of the working face, every 6–8 m Leave 6 × 6 m. A square coal pillar with dimensions of 12 by 6 m is formed on both sides of the coal pillar, leaving 10 m between the neighboring mining sections. The coal pillar is mined in a backward manner. The coal mining method adopted by Sanyi Coal Mine is to dig two 200-m long tunnels on one side of the main tunnel. The drift is 4 m wide, and the section width is 110 m. A connecting tunnel is excavated every 6 m, and 8 m of coal is mined, and 6 m of coal pillar remains.

Strip room-and-pillar mining method: In the sand-based submerged mining area, the overlying bedrock is 20 m. Strips are arranged along both sides of the main tunnel in Daju No. 1 Mine. The method uses five rooms and five pillars with intervals of 5 m along the transport tunnel with a strip width of 6 m and a length of 100 m.

Room-pillar mining method for the opposite working face: Zhaojialiang Mine in the soil-based waterless coal mining area adopts the room-pillar mining method for the opposite working face. The upper and lower transport chute intakes the air, and the middle chute returns the air. The distance between the return air chute and the upper transport chute and the lower transport chute is 70 m. The center distance between the room pillars is 20 m. The room pillar mining width is 16 m, and 4 m is left between the room pillars. The coal pillars are isolated, and the rooms are connected by transport tunnels. A transport tunnel is excavated every 20 m along the direction of the isolated coal pillars, and three rooms are mined simultaneously.

According to many on-site investigations, most of the roofs of the above-mentioned room-and-pillar goafs have not collapsed, and there are many voids in the goafs, which increases the difficulty and workload of goaf management.

4.1.3 Hazard Assessment of Room-and-Pillar Goafs in Coal Mines in Northern Shaanxi

Many room-and-pillar goafs remain underground, posing a great hidden danger to the construction of the overlying ground. Some small coal mines do not have the conditions to be properly sealed and abandoned. The coal in the goafs is connected to the air on the surface, which can easily form underground fire zones, seriously threatening resources and people's safety.

Observations and studies show that the magnitude of surface movement and deformation in the goafs is positively correlated with the mining thickness. Greater surface deformation results from greater mining thickness or more mining layers. The movement of any point on the surface of the goafs must go through the initial period, the active period, and the decay period. The movement amount and speed of each period are different. The sum of the initial, active and declining periods is called the movement duration period. Observations and studies have shown that the length of the movement duration period is related to factors such as the nature of the overburden, mining methods, mining depth and working face advancement speed. The residual movement period can also be called the potential movement period.

In the past, the field of mine surveying and mining subsidence mostly focused on the observation and research during the movement continuation period, especially the dangerous change period. In the formation stage, there is insufficient observation and research on the residual deformation in the recession period, especially after the recession period. It is believed that the residual deformation after the recession period will not cause damage to buildings (structures). However, recent practices have shown that the above view is not entirely correct. Some buildings in goafs where the recession period has long passed have still suffered varying degrees of damage. In the room-and-pillar goafs of small coal mines in northern Shaanxi, due to the low mining rate, although the movement recession period has passed, there are still a certain degree of cracks and cavities in the cross-fracture zone in the upper part of the goaf, which provides conditions for the occurrence and development of residual movement deformation. When the isolation coal pillars in the room-and-pillar goaf are crushed under the long-term compressive stress of the overlying rock strata; or due to long-term weathering and changes in hydrogeological conditions (such as immersion when the water level in the goaf rises, weathering when the water level drops, etc.), the isolation coal pillars between the working faces lose part or all of their supporting function; or under the combined action of the above-mentioned various stresses and the disturbance of other external forces.

The change in the mutual support between the rock blocks across the crack zone in the goaf may lead to a decrease in the crack and void rate in the goaf and its cross-crack zone, which will cause a sudden increase in the residual movement deformation of the surface, thus causing damage to the buildings (structures) on the ground in the goaf. Secondly, in some medium and shallow small coal mine goafs, due to the use of room-and-pillar artificial mining, the mining area is small, the remaining coal pillars are large, and a stress balance arch can be formed in the overlying rock strata. The roof of the goafs can remain suspended for a long time without falling or only partially falling. The surface does not move or moves very slowly within a certain period and does not reach the level of dangerous deformation. Therefore, ground buildings will not be affected by mining for a certain period of time. However, the cavities left underground in such goafs pose potential dangers. After a considerable period of time, the remaining coal pillars in the goaf may partially or completely lose their supporting function due to the long-term compressive stress of the overburden, weathering or other external disturbances, and the internal balance arch of the overburden may be destroyed. The roof and overburden of the goafs may fall off and affect the ground, causing the ground movement and deformation to exceed the allowable value of the building, thereby causing damage to the building.

It can be seen from this that the residual movement (or potential movement) period of the goafs should not be ignored, especially the shallow and medium-shallow goafs of small coal mines should be taken seriously.

4.1.4 Management Approaches of Room-and-Pillar Goafs

The currently used methods for mitigating goafs include filling the goaf with all the support of the overburden; partially supporting the overburden or ground structures to reduce the spatial span of the goaf and prevent the collapse of the roof; strengthening the surrounding rock structure of the goafs by grouting the collapse zone or fracture zone rock mass and enhancing its stability; taking measures to release the settlement potential of the old goafs. However, for such underground voids with a large rate, traditional cement fly ash grouting treatment is difficult to achieve the filling treatment effect, and the slurry has high mobility, resulting in excessive slurry loss rate and increasing the treatment cost.

4.1.4.1 Mitigation Methods

(1) **Full-filling goafs to support overburden**

For goafs with smaller cavities, full filling treatment is currently widely used to eliminate the hidden dangers of surface subsidence. For example, a gypsum goaf in Hunan adopts a filling and grouting scheme. For shallow goafs, direct excavation and stone filling is adopted. For deep goafs, closed walls and gravel and cement grouting are set in the boreholes. The goaf of the coal mine under the Qilin Highway is filled with cement fly ash slurry accelerating coagulation agent. The goaf over Jiaoji railway the section ZH-7 is mainly filled with grouting and sand injection, whereas the goaf over Jinhou Highway is filled with cement fly ash slurry.

(2) **Support overburden at critical locations**

For the goaf that has not collapsed, because the void is too large, the engineering workload and cost of full filling are too high, so local support overburden or ground structures are used to reduce the spatial span of the goaf and prevent the collapse of the roof. Common methods include grouting columns, underground piers, large-diameter bored piles or direct pile foundation method are used for treatment. However, this local filling treatment method has not been widely promoted due to the limitations of goaf conditions, filling materials and technology.

(3) **Grouting reinforcement and strengthening of surrounding rocks**

This method mainly fills the bed-separation voids, fractured zones and deformation formations in the overburden. The objective to create a rock structure with high rigidity and good integrity, which can effectively resist the upward development of the collapse of the old mining area so that relatively balanced subsidence occurs on the surface to ensure the safety of surface structures.

(4) **Taking measures to release the subsidence potential of old goafs**

Before the land overlying the goaf area is used, compulsory measures are taken to accelerate the activation of the old goaf and the subsidence of the overburden. Such engineering measures eliminate the underground cavities that pose a threat to surface safety. The land is then developed and utilized after the subsidence is basically stable. Commonly used methods include top loading and compaction, high-energy level compaction, and water-induced settlement.

In theory, the above methods can be used to manage the underlying goafs. The full-filling management method is the safest and most dependable, but the economic cost is too high. The traditional filling materials are less controllable, resulting in the grouting materials spreading outside the management range, causing waste.

4.1.4.2 Advantages and Disadvantages of Filling Materials

The filling materials used in goaf treatment mainly use cement, clay, fly ash, sand and other materials for filling and reinforcement. If substitute materials are used, they should be used as much as possible, which are low-priced, locally sourced, simple to fill, and have a certain consolidation strength. Currently, the commonly used filling materials are divided into the following types.

(1) **Single liquid cement slurry**

Pure cement slurry or cement as the main agent with a certain quantity of admixtures, water is used to prepare the slurry, and the slurry is injected in a single liquid manner. This slurry is called single liquid cement slurry. Single liquid cement slurry has the advantages of abundant sources, relatively low price, high compressive strength of slurry stone body, good impermeability, simple process equipment, and convenient operation. However, the slurry is a granular material, and it is difficult to inject 0.1 mm or less cracks and sand layer with particle sizes being less than 1.1 mm. In addition, the solidification time is long, and the solidification time is difficult to accurately control.

(2) **Cement clay slurry**

In the single-liquid cement slurry, a certain amount of clay is sometimes added according to the construction purpose and requirements. This slurry is called cement-clay slurry. Compared with the single-liquid cement slurry, cement-clay slurry has lower cost, better mobility and stability, and higher stone formation rate. The addition of clay reduces its compressive strength. This slurry is suitable for filling and grouting.

(3) **Cement fly ash slurry**

In cement slurry, a certain amount of fly ash is added, which is called cement fly ash slurry. Compared with single-liquid cement slurry, cement fly ash slurry has lower cost, better mobility and stability, and higher solidity rate.

The slurry is suitable for filling and grouting. In the grouting and filling project, fly ash is used as a substitute for cement, but it is different from clay. It has a certain activity after mixing with cement and water and has a higher strength in the later stage. In the filling and treatment of goaf areas, water glass is often added to shorten the initial and final setting time of the slurry.

(4) **Water glass slurry**

Water glass can produce gel under the action of acidic curing agent. Water glass itself is abundant in source and cheap. In addition, various new curing agents are constantly being researched and developed, which makes water glass grouts continuously improved. Therefore, this is an increasingly important and widely used grouting material. Due to the wide variety of types, only a few types of grout with more practical value and good effect are introduced in this chapter.

(5) **Other materials**

- Gravel, crushed stone, and others: When the receiving layer is karst and cracks are developed, especially when the groundwater flow rate is high, in order to save grouting materials, sand, gravel, slag, crushed stone, bricks and compressed wood strings and other aggregates are usually poured first to fill large caves and cracks, reduce the water-passing section, increase water flow resistance, reduce groundwater flow rate, and create favorable conditions for grouting. These materials are widely available and inexpensive.
- Cement-Mortar: Generally, cement mortar has a shorter setting time than cement paste and has higher stone strength. The concentration of mortar must be well controlled. If it is too thick, it will be difficult to pump. If it is too thin or the groundwater flow rate is high, the mortar and sand will easily separate.

4.2 Study on Properties of Paste Filling Materials

Paste filling is a cementing filling technology, which is a process of mixing aggregates, cementitious materials, additives and water, stirring and processing them into a toothpaste-like cementing body with good stability, mobility and plasticity, and transporting them to the goafs in the form of plunger flow under the action of gravity or pump pressure to complete the filling operation. Paste filling materials have the characteristics of high solidity, high strength and good economic benefits. They have great advantages for goafs in coal mines where the roof rock strata are relatively complete and hard, the goaf space is relatively large, and the roof has not collapsed or has not completely collapsed.

4.2.1 Selection of Paste Filling Materials

Filling materials can be divided into three categories according to their role in the filling body:

- Inert materials: Inert materials are the main body of filling materials (or aggregates). In filling operations, this type of material is in great demand. It is generally locally available, easy to obtain and cheap. During the filling process and after the filling body is formed, the material properties basically do not change.
- Cementitious materials: In filling projects, cementitious materials will undergo chemical and physical reactions, which can make the inert materials in bulk form cemented bodies with different mechanical properties. In addition, cementitious materials can adjust the filling performance of filling materials, making them more suitable for mixing and transportation. Cementitious materials are also

Table 4.2 Commonly used filling materials

Category	Aggregate (inert material)	Cementitious materials	Adding materials
Material	Waste rock (rock or crushed stone), natural piles, accumulated sand and gravel, crushed sand, tailings, loess, or slag	Cement (including ordinary cement and special cement), finely ground water-quenched slag, pulverized coal ash, lime, or gypsum	Accelerator, water reducing agent, early strength agent, or pumping agent

used in large quantities in filling projects. It is necessary to select more suitable cementitious materials according to the properties of inert materials. The cost of cementitious materials is the main component of the cost of filling project materials.

- Additives: Additives are additives used to improve certain properties of filling materials or to improve the quality of filling. Generally, the dosage used is small, but the price is more expensive.

Commonly used filling materials are presented in Table 4.2.

(1) **Filling aggregate**

The selection of filling materials should be based on the principle of local materials. The materials must be sufficient and easy to collect, process and transport. A large amount of aeolian sand is enriched on the surface of the southern wing of Ningtiaota Coal Mine, which can be used as the main filling material for filling the goaf. The aeolian sand material needs to be screened to remove impurities. Aeolian sand is the main material in the filling material and plays the role of aggregate.

(2) **Filling cementitious material**

Among the commonly used filling methods, cement, fly ash, lime and gypsum are commonly used cementing materials. According to the characteristics of cementing materials, cement and fly ash are selected as cementing materials. Ordinary Portland cement and fly ash use secondary fly ash dry ash.

4.2.2 Raw Material Analysis

The raw materials include cement, fly ash, and aeolian sand.

(1) **Cement**

Qinling PO42.5 cement is commonly used. The density is 3.1 t/m^3. The main chemical composition is summarized in Table 4.3.

The cement particle size distribution measured by laser particle size analyzer is shown in Fig. 4.1.

Table 4.3 Cement chemical composition analysis

Sample	Main chemical compositions (%)						
	SiO_2	Al_2O_3	Fe_2O_3	MgO	CaO	Na_2O	K_2O
Cement	19.85	4.95	2.63	2.57	59.67	0.35	0.84

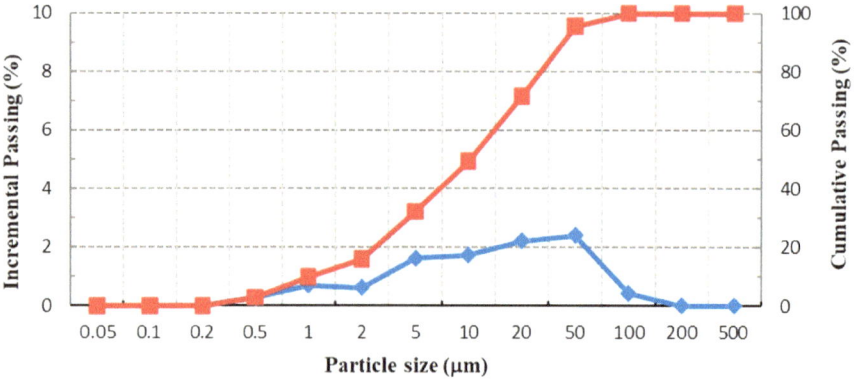

Fig. 4.1 Particle size distribution of Qinling PO42.5 cement

(2) **Fly ash**

The test fly ash was taken from Shenmu Jinjie Power Plant. It is secondary ash with a bulk density of 0.92 t/m^3 and an apparent density of 2.22 t/m^3. The retaining rate passing through the 0.045 mm sieve is approximately 36%. Its main chemical composition is presented in Table 4.4.

The fly ash particle size distribution measured by laser particle size analyzer is shown in Fig. 4.2.

(3) **Aeolian sand**

The test aeolian sand was taken from Yulin Yuyang District, with a bulk density of 1.53 t/m^3 and an apparent density of 2.60 t/m^3. The porosity of sample is 41.15%, and the water content is 4.05%. The sand contains impurity material with a mud content of 1.76%. Its main chemical components are shown in Table 4.5.

The particle size distribution of aeolian sand measured by laser particle size analyzer is shown in Fig. 4.3.

Table 4.4 Chemical composition analysis of fly ash

Sample ·	Chemical compositions (%)						
	SiO_2	Al_2O_3	Fe_2O_3	MgO	CaO	Na_2O	K_2O
Fly ash	53.58	25.55	5.71	1.27	6.43	0.57	1.37

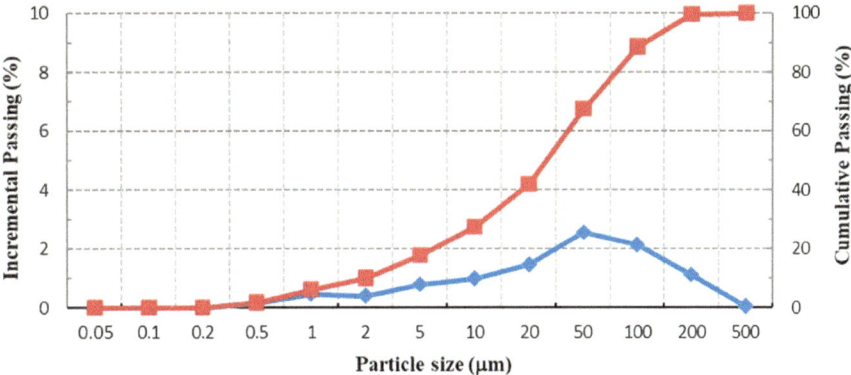

Fig. 4.2 Particle size distribution of fly ash

Table 4.5 Chemical composition analysis of aeolian sand

Sample	Chemical compositions (%)						
	SiO_2	Al_2O_3	Fe_2O_3	MgO	CaO	Na_2O	K_2O
Aeolian sand	78.14	12.35	0.96	0.23	1.22	3.26	2.99

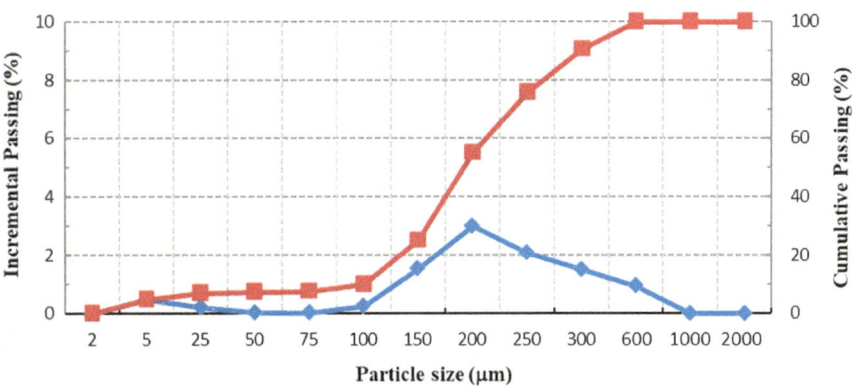

Fig. 4.3 Particle size distribution of aeolian sand

4.2.3 Test and Characteristics of Aeolian Sand and Paste Ratio

4.2.3.1 Aeolian Sand and Paste Parameter Requirements

Paste filling materials are made by mixing aeolian sand, fly ash, cement, additives and water in a certain ratio. To ensure the transportation and performance of the filling materials after filling, the prepared filling materials must meet the following specific conditions:

- Initial setting time: The initial setting time of the filling material should be longer than the time required for the entire process of slurry transportation, pumping, and pouring, and the initial setting time should be in the range of 4–6 h.
- Collapsibility: The high-concentration materials in the mix must reach a certain height before collapse to meet the requirements of pipeline pumping in the filling system. Usually, the collapsibility of the slurry needs to reach a range between 20 and 25 cm.
- Particle grading: The maximum particle size of the ingredients should not be greater than 25 mm, and the particle grading should adopt continuous grading. The ultra-fine particle content should not be less than 15%, and when the stability requirement is high, the particle content finer than -20 μm should not be less than 25%.
- Mobility: Since the only pumpable substance in the raw materials of the filling material is water, and the water content under the pumping pressure controls the pumping performance, the mobility is related to the degree of stratification, which is less than 2.
- Water retention: During the pumping process, the lubricating layer composed of water, cement and fly ash plays a lubricating role between the filling material and the pipe wall. When the pumping pressure exceeds the friction resistance of the pipe wall, the filling material moves forward along the conveying pipe. If the filling material has good mobility but poor water retention, water seepage and stratification will occur in the conveying pipe. When water floats to the surface and aggregates are deposited, the pipe will be blocked or burst. Therefore, water retention is an important indicator of the conveying performance of the filling material, and the water seepage rate of the material is required to be 10% lower.
- Strength: After the paste forms a filling body, it must have a high late strength and a low compression rate to play the role of a bearing structure, to give full play to the effect of filling to reduce surface subsidence and ensure the safety of surface buildings and. According to the requirements of the specification "Technical Specifications for Design and Construction of Goaf Highways", the single compressive strength of the paste should not be less than 2 MPa.

4.2.3.2 Material Ratio Tests

The key to the mix ratio of aeolian sand paste filling is to determine the specific required strength, setting time, self-flow or pumping performance and water seepage rate and other technical indicators according to the filling conditions, and to select the material ratio and concentration that are acceptable in terms of technology, economy and engineering in combination with the material engineering cost.

Experiments were conducted according to different ratios, and the ratios that meet the performance requirements such as mobility, stability, water retention rate and strength were obtained. Table 4.6 presents the materials used for the tests.

Table 4.6   Material used in tests

Element	Aeolian sand	Fly ash	Cement	Water
Content (kg/m^3)	1419	327	109	244

(1)   Uniaxial compressive strength

The main instruments and equipment used for the tests are WEP-600 computer-controlled screen display universal testing machines. The uniaxial compressive strength is calculated by:

$$\sigma_c = P_{max}/A$$

where

σ_c uniaxial compressive strength of filling material (MPa)
P_{max} maximum failure load of filling material specimen (kN)
A   compressive area of specimen (mm^2).

The compressive strength test results are shown in Table 4.7.

The experimental results suggest that the average uniaxial compressive strength of the aeolian sand and paste material can reach 0.93 MPa in three days. The 7-day uniaxial compressive strength reaches 1.90 MPa. The 14-day uniaxial compressive strength reaches 3.12 MPa, which can meet the strength requirements of the filling material in the goaf area.

Table 4.7   Test results of compressive strength

Serial number	Density (g cm^{-3})	Load (kN)	Uniaxial compressive strength (MPa)	Maintenance time (d)
3-1	2.11	4.51	0.92	3
3-2	2.08	4.79	0.95	3
3-3	2.09	4.56	0.91	3
Average	2.09	4.52	0.93	3
7-1	2.11	9.45	1.85	7
7-2	2.10	9.00	1.89	7
7-3	2.11	9.65	1.98	7
Average	2.11	9.37	1.90	7
14-1	2.08	16.00	3.22	14
14-2	2.12	15.22	3.02	14
14-3	2.07	15.28	3.12	14
Average	2.09	15.50	3.12	14

(2) **Mobility performance**

There are many parameters that affect transport of paste materials including collapsibility, degree of stratification, water bleeding rate,

Collapsibility: The main instruments and equipment for collapsibility tests include collapsibility cone, tamping rod, slump gauge, trowel, and small shovel. The testing method includes the following:

- Mix the concrete mixture as required and use a shovel to pass the concrete mixture through the bucket funnel. Then divide it into 3 layers and load into the cylinder with each layer having a roughly equal volume. The bottom layer is about 70 mm thick, and the middle layer is about 90 mm thick. Each time a layer is loaded, the cylinder is tamped evenly 25 times with a tamping rod.
- After the upper layer is tamped, remove the charging funnel, smooth the mouth of the tube with a trowel, and then slowly lift the slump barrel vertically. When the sample no longer collapses, use a steel ruler to measure the height difference between the center point of the top of the sample and the slump barrel, accurate to 1 mm.

Degree of stratification: The main instruments and equipment used for stratification tests include mortar consistency meter, mortar stratification meter, stopwatch, mixing pot, metal tamping rod, sharpening knife, and wooden hammer. The Testing methods include the following:

- Carry out a consistency test on the mixed mortar first, and then fill the stratification cylinder with the same batch of mortar (or remix the mortar that has undergone the consistency test) at one time.
- After standing for 30 min, remove the upper 200 mm mortar, then take out the bottom 100 mm mortar, re-mix in the mortar mixing pot, mix for 2 min, and then measure the mortar consistency value.
- The difference between the two mortar consistencies is the mortar stratification value (in mm).
- The mortar delamination degree should be the arithmetic means of the two test results. The test should be repeated if the difference is greater than 20 mm.

Water leaching rate: Main instruments and equipment used for water leaching rate include measuring cylinder, vibrating table, measuring cylinder, metal tamper, and pipette. The water leaching rate is calculated by the following equation:

$$B_m = \frac{W_b G}{W G_1} 100\%$$

where

B_m leaching rate (%)
W_b total mass of water leached (g)

Table 4.8 Test results of mobility performance

Series number	Mass fraction (%)	Collapsibility (cm)	Consistency (cm)	Degree of stratification (cm)	Water leachate rate (%)
1	85.5	21	9.6	1.1	2.34
2	85.0	21	10.1	1.3	2.55
3	84.7	25	10.5	1.2	2.40
4	83.1	24	10.2	1.4	2.78

W   total amount of water used in one mixing (g)
G   total mass of mortar mixed in one mixing (g)
G1   mass of sample (g).

The test results are summarized in Table 4.8.

The experimental results suggest that the collapsibility, degree of stratification, and water leachate rate of aeolian sand and paste materials all meet the material requirements of transportation performance.

4.3 Goaf Remediation in Ningtiaota Mine

4.3.1 Project Overview

The industrial square of the northern wing of Ningtiaota Coal Mine is in the north of Kaokausu Valley and is currently under construction. On September 12, 2013, the mortar-laid rubble slope protection around the site cracked and deformed, and a large area of the site collapsed, deformed and cracked. As of September 17, 2013, the maximum subsidence amount of the coal tower in the site reached 790 mm, and the maximum tilting value reached 240 mm. The maximum sinking amount of the screening workshop reached 155 mm, and the maximum tilting value reached 237 mm. Cracks appeared in the main structures. Preliminary investigations indicate that the subsidence and surface deformation are caused by the collapse of the underground goafs. To reduce the damage to the ground engineering caused by the continuous collapse of the underground goafs and eliminate the disaster risks of the goaf. An emergency rescue and remediation of underground mining collapse disasters is conducted with grouting and back filling on key parts of the underground goaf area of the site.

4.3.1.1 Topography

The northern wing of the mine is located to the north of Kaokauwusu Valley, which is a loess hilly gully area. The surface loess and red soil cover the bedrock with a large thickness, generally 20–100 m. Due to the action of external forces, a series of special loess landforms are formed, with complex terrain, crisscrossed valleys, alternating ridges and hills, fragmented terrain, steep and narrow valleys. The surface erosion is strong, with short gullies of varying densities. Modern landforms are mainly caused by water erosion, with sparse vegetation, severe soil and water loss, and bedrock exposed on both sides of the valley.

There are loess ridges on the east and west sides of the treatment area. The thickness of the loess on the west side is about 20 m, and the bedrock has been exposed after the site is leveled; the thickness of the loess on the east side is relatively large, and the bedrock has not been exposed after the site is leveled. In general, the North Wing Measure Well Industrial Site is in a valley sandwiched between two ridges.

4.3.1.2 Stratigraphic Lithology

According to previous data and the drilling revelations, the strata in the treatment area are as follows from old to new: the Yan'an Formation of the Middle Jurassic System (J2y), the Zhiluo Formation (J2z) and the Quaternary System (Q). They are described from old to new as follows:

Middle Jurassic Yan'an Formation (J2y)

This group of strata is pseudo-integrated on the underlying Triassic Upper Yongping Formation, and is the coal-bearing stratum in this area, with a thickness of 170.52–240.9 m and an average of 208.97 m. The upper part has been eroded to varying degrees. In the exploration area, this group of strata gradually thins from the middle to the surrounding areas. Most of it is covered by the overlying strata, and only the upper strata of this group are intermittently exposed in the valleys of Kaokauwusu Valley and Kentieling River. This group of strata is a set of terrigenous clastic deposits, and the lithology is composed of light grayish white medium-fine-grained feldspar sandstone, lithic feldspar sandstone, gray to blue-gray sandy mudstone, mudstone and coal seams, with a small amount of calcareous sandstone, carbonaceous mudstone, etc.

Middle Jurassic Zhiluo Formation (J2z)

Due to the late denudation, only the lower part of the Zhiluo Formation remains in the area. The upper part of the formation is yellow-green sandy mudstone and siltstone, containing siderite nodules, and the lower part is gray-green thick-bedded medium-coarse feldspar sandstone, intercalated with gray mudstone, which is yellow green after weathering. Some drilling holes reveal purple variegated patches in the upper part of this formation. It has large-scale plate-like crossbedding or no obvious bedding, containing plant stem and leaf fossils, mirror coal lumps and pyrite nodules. The bottom sandstone occasionally contains quartz gravel, with gravel diameters ranging from 2 to 150 mm, forming the bottom conglomerate of this formation, which is easy to distinguish from the Yan'an Formation and is in unconformity contact with the underlying coal-bearing stratum Yan'an Formation.

Upper Pleistocene Malan Formation (Q3m)

This group of strata is distributed in Liangmao area in the north of the well field and in the southwest of the well field. The thickness of this group of strata in the treatment area is 0–10.0 m, with an average of 4.90 m. The lithology is grayish yellow sub-sand soil, which feels rough when touched by hand. It contains a small amount of scattered calcareous nodules, with developed vertical joints, forming steep walls, relatively uniform lithology, loose structure, and large pores.

4.3.1.3 Geological Structure

Ningtiaota Coalfield is located I the broad and gentle eastern wing of the Ordos syncline of the northern Shaanxi slope. The strata of each era (except the Cenozoic) are in an integrated or pseudo-integrated contact relationship. The surface outcrops of the coal-bearing strata are nearly horizontal, and the underlying coal seam floor is inclined to the west with a slope of 4–10‰. In general, it is a monocline layer inclined to the west, reflecting the characteristics of the regional structure. No drop greater than 15 m was found in previous geological surveys. There are no faults, no signs of magmatic activity, and the structure is simple. However, judging from the contour lines of the coal seam floor, there are some wide and gentle undulating phenomena.

4.3.1.4 Hydrogeological Conditions

Porous-Media Aquifer of Upper Pleistocene Malan Formation (Q_{3m})

The surface of the northern wing mining area is continuously distributed, with a thickness of 0–10 m, generally 4.9 m. The lithology is light yellow or grayish yellow sub-sand soil, with uniform texture, developed vertical joints, loose structure, large pores, and a weakly permeable layer.

Fractured Aquifer of Yan'an Formation (J2y)

Coal seam 2^{-2} is directly filled with water. The thickness of the section ranges from 23.70 to 43.70 m, with an average of 35.27 m. The thickness of the aquifer is 6–37 m, typically 24 m. It is mainly composed of fractured aquifer that is mostly under confined conditions. Some parts are phreatic, but the water content is weak. The unit water yield is 0.0000652–0.000853 L/s m. The hydraulic conductivity varies from 0.000269 to 0.004356 m/d. The water chemistry type is HCO_3-Na·Ca.

4.3.1.5 Coal Seams

According to the existing data, the main coal seams in the mine field are coal seams $1^{-2}, 2^{-2}$, and 3^{-1}. The characteristics of each coal seam are summarized in Table 4.9.

Coal seam 1^{-2}: Coal seam 1^{-2} is located in the middle and upper part of the fifth section of the Yan'an Formation. It is distributed to the north and south of Kaokauwusu Valley and is divided into 5 irregularly shaped isolated recoverable blocks. Three blocks in the mining area are small. Two contiguous blocks are large, one of which is distributed west of the Kentieling River, and the other is distributed in the north of Kaokauwusu Valley. The largest contiguous area and a mineable area is 57.156 km^2. The coal seam thickness ranges from 0.00 to 2.34 m, with an average coal thickness of 1.14 m. The coal seam thickness within the mineable range varies from 0.80 to 2.34 m, with a standard deviation of 1.05 and a coefficient of variation of 0.45. The coal seam contains 1–2 gangues. The gangue composition includes mudstone and siltstone. The coal seam is buried at a depth between 0 and 225.66 m, and the floor elevation ranges from 1106 to 1196 m above mean sea level.

This coal seam is thin to medium thick. The coal seam in most areas is mineable. The thickness of the coal seam within the mineable area varies little and has obvious regularity. The structure is simple. The coal type is single. The ash and sulfur contents of the coal seam remain low and persistent. It is a relatively stable coal seam.

Coal seam 2^{-2}: This coal seam is located at the top of the fourth section of the Yan'an Formation and is the main mineable coal seam. The mineable area is 111.828 km^2. The coal seam thickness ranges from 0.70 to 9.46 m, and the average thickness is 6.11 m. The coal seam gradually becomes thinner from east to west and from south to north. Generally, the thickness of the coal seam varies from 4 to 5 m with 6.50 to 7.50 m in the east. The coefficient of variation is 0.24. It is a single coal seam without interlayer or with one interlayer gangue. Some places contain 2–3 layers of gangue. The gangue is 0.04–0.36 m thick. The lithology is mudstone, carbonaceous mudstone and siltstone. The roof of the coal seam is made of gray siltstone, and locally grayish white medium-grained sandstone. The coal seam is buried at depths between 2.00 and 247.01 m. The floor elevation is 1088–1170 m above mean sea level. This coal seam has outcrops near Kaokauwusu Valley. Many small coal mines mine this coal seam.

The remediation area is associated with coal seam 2^{-2}. The coal burial depth is between 51.7 and 54.0 m. The coal seam bottom elevation ranges from 1127 to 1130 m above mean sea level. The coal seam thickness varies from 5.6 to 6.5 m.

Coal seam 3^{-1}: Located at the top of the third section of the Yan'an Formation, this coal seam spontaneously combusts in the Kaokauwusu Valley area and Liushuihao area in the middle and eastern part of the mine field (the spontaneous combustion area is 3.6 km^2). The mineable area is 129.628 km^2. The coal seam gradually becomes thinner from south to north. The coal seam thickness is between 1.82 and 3.24 m, with an average thickness of 2.85 m and a coefficient of variation of 0.04. It generally does not contain interlayers but contains one layer of gangue. The thickness of the

Table 4.9 Main characteristics of each coal seam

Coal seam	Stratum	Thickness (m) $\frac{Min - Max}{Average}$	Elevation of coal seam bottom (m)	Average depth (m)	Separation distance (m) $\frac{Min - Max}{Average}$	Mineable rate (%)	Stability	Mineable evaluation
1^{-2}	Middle of fifth section	$\frac{0 - 2.34}{1.14}$	1175.64	16.4	8.14–44.03	54	Relatively stable	Most of the coal seams can be mined
2^{-2}	Top of fourth section	$\frac{0.70 - 9.46}{6.11}$	1132.91	52.1	$\frac{22.73-52.5}{28.40}$	100	Stable	
3^{-1}	Top of third section	$\frac{1.82 - 3.24}{2.85}$	1102.39	89.8	33.19	100		

gangue is between 0.02 and 0.22 m. The rock types of the intercalated gangue are mudstone and siltstone. The roof of the coal seam is generally composed of sandy mudstone or siltstone, and locally fine-medium grained sandstone. The coal seam is buried at depths ranging from 33.86 to 287.54 m. The floor elevation varies from 1049 to 1135 m above mean sea level.

4.3.2 Mining Characteristics

The room-and-pillar goafs in the project area are shallowly buried, approximately 55 m. The residual coal pillars bifurcate over time. The coal pillar walls start peeling off and are plastically deformed under the long-term mine-induced pressures. With the decrease of the effective residual coal pillar rate the roof sags under the gravity of the overlying rock layer, causing the goaf to become unstable and thus threatening the safety of ground buildings (structures). This type of goaf roof that is supported by residual coal pillars is not stable, and its stability is controlled by highly nonlinear characteristics. A small disturbance of a control variable can cause the system to lose equilibrium between forces, resulting in sudden changes in stress state and collapses on surface. Figure 4.4 shows the various factors that affect stability of goafs.

Investigation of the mining conditions indicates that the underlying 2^{-2} goafs are caused by former Shaqu Coal Mine. The coal seam that is mined has a thickness of approximately 6 m and a burial depth of approximately 55 m. Shaqu Coal Mine operated for recovery mining from May 2007 to May 2008. Nine rooms were mined while seven pillars were kept intact. Its mining characteristics include the use of short working faces to advance, while the use of coal pillars is either temporary or

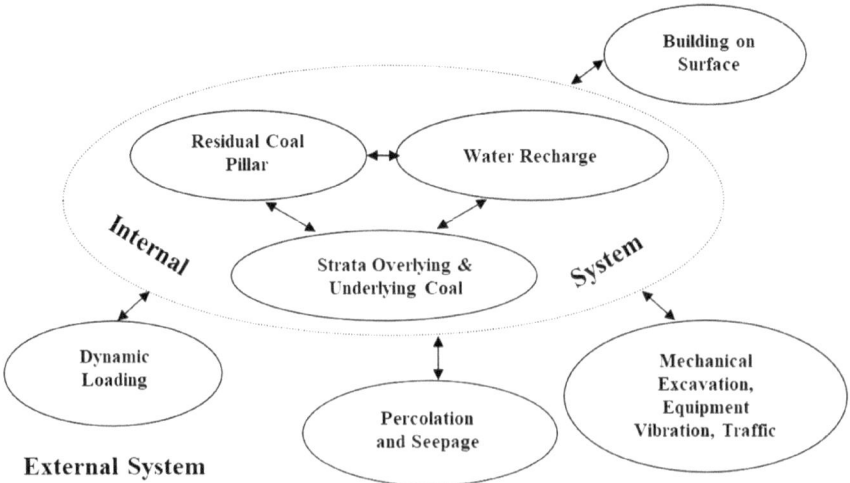

Fig. 4.4 Factors affecting stability of old empty areas over room-and-pillar mining areas

permanent. The coal pillar sizes vary with strong irregularity. The roof overlying the coal seams allows to cave.

The closed Shaqu Coal Mine has a mining system with tunnels on the same level. The main tunnels and auxiliary tunnels for the panel areas are parallel and horizontal. According to the layout of the main tunnels, auxiliary tunnels in panel area are arranged on both sides of the main tunnel. The panel area coal room is arranged on both sides of the panel area auxiliary roadway. The rooms are the mining area, which is 7 m by 7 m. The blocky coal pillars are recovered in a planned manner according to the site conditions after the room mining. The mining data of this area was not preserved, and there are no as-built plans. The existing data only provides the general location of the mining boundary of the coal mine. Since the recovery of coal pillars during room-and-pillar mining is highly variable in actual operations, the actual distribution of the goafs is unclear, which brings great difficulties to the management of the goafs. According to the results of the emergency management engineering survey of the goaf collapse in the mining area and general knowledge of the characteristics of room-and-pillar mining, the distribution of the underlying goaf of the industrial site can be preliminarily analyzed as shown in Fig. 4.5.

Fig. 4.5 Distribution of underground goafs

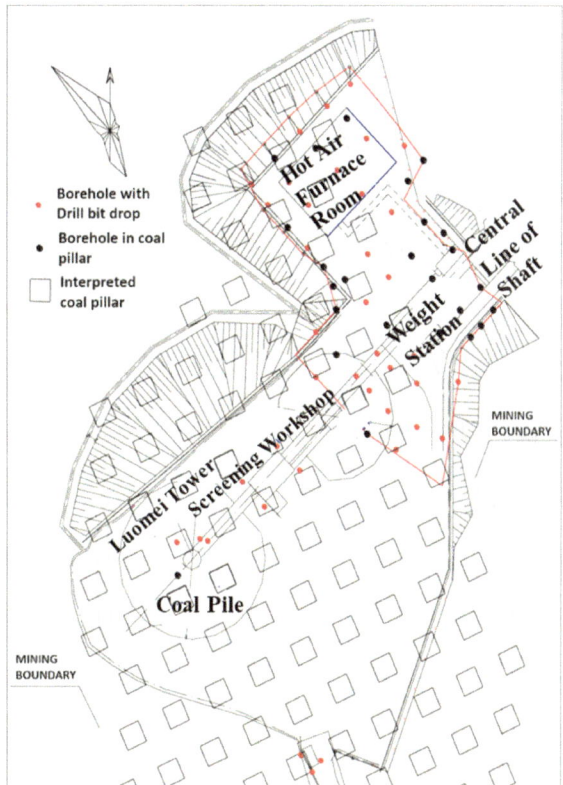

4.3.3 Comparison and Selection of Mitigation Approaches

Many mitigation methods are available for goaf mitigation in China, including filling method, local support method and release of settlement potential method. The goafs of the northern wing of the Ningtiaota Mine are shallow goafs. According to the experience of goaf treatment in northern Shaanxi and the current conditions of the exploration area, the goaf remediation adopts the intermittent grouting scheme.

The remediation area includes the wellhead, drive workshop, trestle and hot air furnace room exploration area in the North Wing. The coal burial depth ranges from 52.0 to 57.0 m. The goaf management width is 30 m outside the foundation. In the hot air furnace room and pipeline areas, the enclosure width is 10 m. The total remediation area is 5500 m^2. According to the survey results of the emergency treatment project, the goaf height is 5 m. The recovery rate is 70%, while the remaining deformation is 95%.

Estimation of Goaf Volume

The volume of coal mine goafs is difficult to calculate accurately. According to the Technical Specifications for Design and Construction of Highway over Goafs (JTG/T D31-03-2011), the formula for estimating the remaining void volume in the goaf is:

$$V_{void} = S \times M \times K \times \Delta V$$

where

S   goaf area (m^2)
M   coal seam mining height (m)
K   recovery rate (%)
ΔV Residual porosity (%).

According to the equation above, the remaining void volume in the goafs is approximately 18,288 m^3.

Borehole Layout

The remediation adopts the drilling and grouting method to fill the goafs. Drilling is conducted in the industrial site of the measure well. The final borehole level is 2^{-2} coal seam, while using grout borehole to fill and grout the goafs. Forty-seven grouting boreholes are arranged in the wellhead and hot air furnace room remediation area. Four inspection boreholes are designed in each of the four areas: Luomei Tower, screening workshop, and two weight stations, for a total of 16 boreholes. The specifications of the boreholes are summarized Table 4.10.

4.3.3.1   Traditional Filling Solution

The grouting material consists of grouting slurry and coarse aggregate.

Table 4.10 Summary of borehole specifications

Remediation area	Number of boreholes	Footage (m)	Quaternary thickness (m)	Bedrock thickness (m)
Wellhead and hot air furnace room	47	3055	376	2679
Luomei tower, screening workshop	8	480	40	440
Weight stations	8	480	40	440
Total	63	4015	456	3559

(1) The slurry is cement fly ash slurry, and the slurry has following ratios:

- Slurry: water: solid phase ratio: 1:1–1:1.2
- Cement: fly ash mix ratio: 4:6 for grouting holes; 3:7 for curtain grouting holes
- Accelerator: add 2% of the weight of cement to the holes drilled in the tunnels. When the grouting volume is large in the grouting holes of structures and general grouting holes, add 2% of the weight of cement to the slurry.

(2) When the drill bit suddenly dropped during drilling, and the drill bit dropped more than 0.5 m, coarse aggregates are added to the slurry. In principle, the coarse aggregate should not exceed 30% of the total grouting volume.

Before construction, the grouting material ratio should be determined by laboratory and field tests according to the cement and fly ash used during construction (ratios are 1:1, 1:1.1, 1:1.2, respectively). The testing parameters include dry material content per cubic meter of slurry, slurry density, initial and final setting time, grout consolidation rate, and strength of consolidated grout.

Cement fly ash slurry has high mobility. The slurry consumption coefficient, A, is assumed to be 0.3. The grouting project quantity is presented in Table 4.11.

Table 4.11 Goaf area grouting volume estimation with cement fly ash slurry

Remaining void volume (m^3)	Slurry consumption coefficient, A	Filling coefficient, η	Grout consolidation rate, C	Grouting (sand) quantity (m^3)	Slurry volume (m^3)	Sand volume (15% of the total grouting and sand volume) (m^3)
18,288	0.3	0.85	0.7	28,869	24,539	4330

Table 4.12 Goaf area grouting volume estimation with paste slurry addition

Remaining void volume (m^3)	Slurry consumption coefficient, A	Filling coefficient, η	Grout consolidation rate, C	Total filling volume (m^3)	Cement fly ash slurry quantity (m^3)	Paste slurry volume (m^3)
18,288	0.25	0.85	0.75	25,908	5182	20,726

4.3.3.2 Paste Material Solution

The grouting material consists of paste slurry and cement fly ash slurry.
Paste to slurry ratio is as follows:

- aeolian sand: cement: fly ash = 1:3–1:3
- cement fly ash slurry: water = 1:1–1:1.2.

The grouting boreholes are filled with paste slurry in curtain borehole that has drill bit drop greater than 0.5 m. Other grouting boreholes are filled with cement fly ash slurry. When injecting cement fly ash slurry, 2% of the weight of cement is added to the curtain grouting boreholes, and 2% of the weight of cement is added to the slurry when the grouting amount is large for the grouting boreholes under the foundation of the structure and the general grouting holes.

The average slurry consumption coefficient of paste slurry and cement—fly ash slurry is A = 0.25, and the average grout consolidation rate is C = 75%. The grouting project quantity is shown in Table 4.12.

4.3.3.3 Technical and Economic Comparison

Comparison of Technical Conditions

The two treatment schemes are consistent in terms of filling range and drilling layout, and the main differences lie in the filling materials and filling processes:

The traditional filling scheme uses cement fly ash slurry filling. Where the voids are larger than 0.5 m, sand filling is also required in the boreholes. This process technology is relatively mature, with a total of slurry preparation, mixing, grouting and sanding as the key construction points. Among them, on-site slurry preparation is difficult to accurately prepare the slurry according to the designed proportion due to the limitation of conditions, and it is necessary to frequently debug the slurry preparation equipment and test the slurry proportion; sanding construction is limited by the borehole diameter, and the amount of sand added is small, which can only reduce the amount of cement fly ash slurry to a small extent, and the slurry and sand cannot be fully mixed in the borehole, resulting in uneven stone body, and the compressive strength is difficult to meet the design requirements.

Paste material filling is a new method of filling and remediating goafs. Paste material is directly mixed by concrete mixing stations, with low material mobility, high consolidation rate, and good controllability of filling range. The filling process of paste materials is often divided into two types: pouring and pressing. Both are simple to operate.

Economic Comparison

Of the two grouting schemes, the traditional filling remediation method requires a large volume of grout, and the cement fly ash slurry has a low consolidation rate. It is difficult to ensure that the grouting body is completely connected to the top when filling in the cavity. The construction process is complicated, the mitigation effect is difficult to meet the design requirements. Furthermore, the construction cost is too high. Paste materials have low mobility, high consolidation rate, good filling controllability, and simple construction process. The method using paste materials can reduce construction costs to the greatest extent while meeting the design requirements.

4.4 Evaluation of Remediation Effectiveness

The grouting remediation of the goafs is concealed because the filling processes occur underground. Its quality control has always been a challenge. It is difficult to verify the quality of the grouting effect with a single detection method. Only by adopting comprehensive technology with multiple detection methods can satisfactory results be achieved. The technologies and methods used for engineering quality inspection mainly include geophysical prospecting, drilling coring and laboratory testing, borehole water pressure testing, tracer testing, and surface deformation observation. The inspection time should be conducted six months after the completion of the grouting construction. Considering that the grouting pressure will produce a large stress change in the underground goafs in response to the grouting, a certain time is allowed for compressive stress adjustment in the goaf to achieve the balance and stability of the goaf foundation after the grouting is completed.

4.4.1 Construction Status

The remediation project started on March 26, 2014, and was completed on May 17, 2024. A total of 69 boreholes were completed with the footage of 4611 m. Tables 4.13 and 4.14 summarize the works completed. The total grouting volume is 28,226 m^3, including 21,931 m^3 of paste and 6295 m^3 of cement fly ash slurry.

The remediation adapts a technology the involves one-time borehole-advancing and one-time full-hole intermittent grouting and sand filling from bottom up. The grout consolidation rate is greater than 75%. Boreholes are drilled on the surface

Table 4.13 Summary of drilling tasks

Locations	Designed footage (m)/number of boreholes	Completed footage (m)/number of boreholes
Wellhead and hot air furnace room	3055/47	3321/53
Luomei tower, screening workshop	480/8	487/8
Weight stations	480/8	489/8

Table 4.14 Summary of filling tasks

Locations	Designed volume (m³)	Completed volume (m³)	Remark
Wellhead and hot air furnace room (paste volume/cement fly ash slurry volume)	20,726/5182	19,793/5665	Sand injection 2411 m³
Luomei tower, screening workshop (paste volume/cement fly ash slurry volume)		981/580	
Weight stations (paste volume/cement fly ash slurry volume)		1157/50	

and paste slurry or cement fly ash slurry is injected into the goafs, collapse areas, and the cracks of the overlying rock mass (goaf collapse) through concrete pumps or grouting pumps and grouting pipes. The slurry is solidified and cements the fractured zones in the rock strata. The stone body formed by the slurry in the goafs or collapse areas supports the overlying rock strata to ensure the stability of the site foundation.

4.4.2 Evaluation Method

Verification of the quality of the grouting effect has been a challenge. Definitive conclusions need multiple detection methods, as discussed below.

(1) **Geophysical exploration**

Geophysical detection technology is an important method for inspecting the quality of goaf control projects after they are completed. It compares the changes in the physical properties of the rock formations in the same range, point, and depth in the goaf area before and after grouting to intuitively judge the quality of the project. Its advantages are low cost, fast speed, high efficiency, and simple construction. However, due to the multi-solution nature of geophysical data, only qualitative evaluation of the project quality can be performed. Commonly used methods include in-hole wave velocity logging, Rayleigh wave (surface wave) method, transient electromagnetic method, or high-density electrical method.

(2) **Drilling and coring and laboratory testing**

Drilling and coring are the main technologies and methods in the quality inspection of goaf treatment projects and can provide a working platform for in-hole geophysical detection and water pressure test. The degree of cementation between the slurry stone body and the surrounding rock can be judged according to the drilling core sampling rate and the degree of core crushing. The filling and cementation degree of the slurry on the broken rock layer can be judged according to the consumption of circulating fluid and the water pressure in the hole during the drilling process. By coring the slurry stone body, the final solidification degree of the slurry in the underground can be understood, and the stone body can be subjected to indoor compressive strength test to check whether its strength meets the design requirements.

(3) **Water pressure test**

Representative boreholes can be selected for comparison of water pressure tests before and after grouting, and the quality of goaf control projects was checked based on the change in water intake per unit length.

(4) **Surface deformation observation**

Whether the foundation stability after goaf treatment can meet the requirements of highway engineering is mainly evaluated through surface deformation observation. This method is intuitive and highly accurate, but the working time is long, and the location selection and protection of the measuring points are difficult. Once the measuring points are damaged, the observation work will be wasted. Therefore, surface deformation observation is currently only used in some goaf treatment projects, but the actual observation effect plays a vital role in the evaluation of treatment projects.

4.4.3 Surface Deformation Monitoring Results

At completion of the mine shaft in Ningtiaota Coal Mine, cracks were reported in the mountains and ground around the site. To ensure the safe operation of the mine, deformation monitoring is conducted in the coal bunker, Luomei Tower, and screening workshop buildings. Figure 4.6 shows the monitoring stations, which include 2 monument stations, 3 leveling points, and 10 subsidence monitoring points. Monitoring points S1 through S5 are in the screening workshop area. Monitoring points Y1 through Y3 are in the Luomei Tower, and monitoring points Z1 and Z2 3 are in the coal bunker area.

Figures 4.7, 4.8, 4.9, 4.10 and 4.11 show the subsidence data at S1 through S6. The elevations at S1, S2 and S3 demonstrate approximately 1 mm rise because of

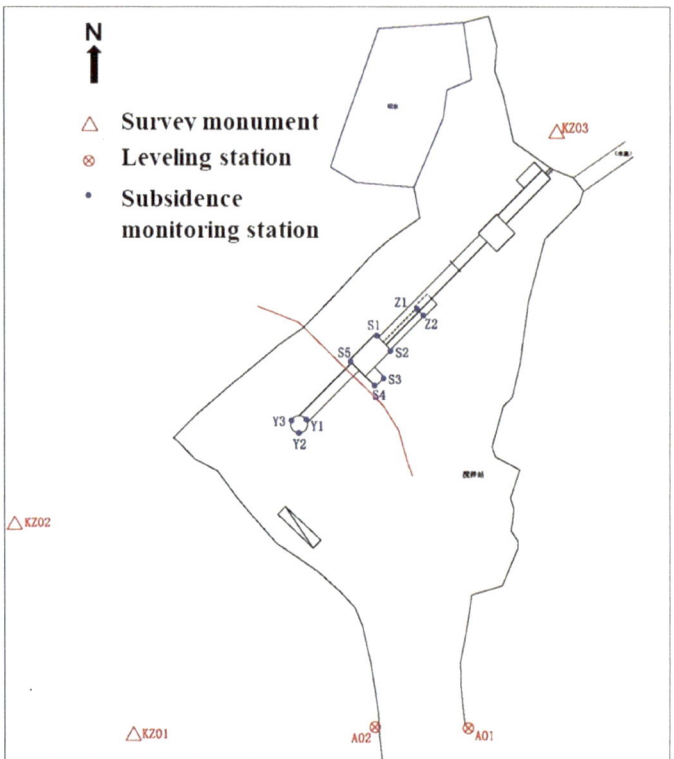

Fig. 4.6 Measuring point layout

the grouting and filling remediation. The ground elevation increase indicates that the paste material has effectively filled the previously unfilled voids and played a certain role in correcting the deviation of the building. The data at S4 and S5 tends to be stable.

Figures 4.12 and 4.14 show the subsidence data at monitoring stations Y1 though Y3. The subsidence appears to stop two months after the completion of the remediation. The elevations of the monitoring points stopped settling, suggesting that the paste material may have effectively filled the voids in the goafs.

Figures 4.15 and 4.16 show the subsidence data at monitoring stations Z1 and Z2. The elevation f at the monitoring point increased by 1.7 mm in response to the remediation. The elevation became stable two months after the remediation was completed. Essentially, the buildings stopped settling, and the paste material had effectively filled the voids in the goafs.

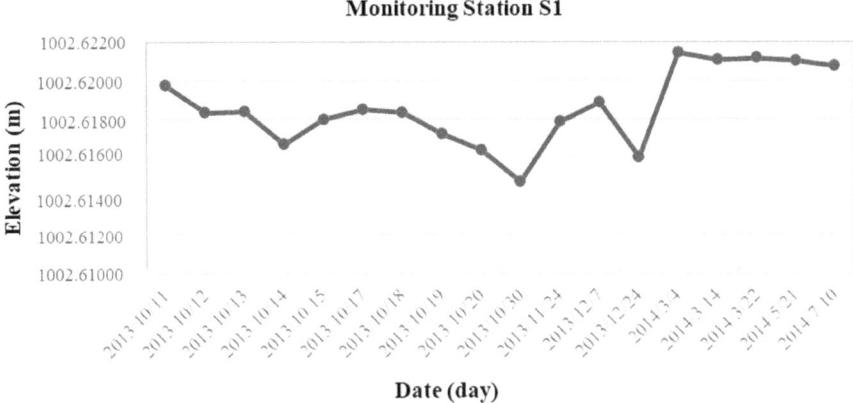

Fig. 4.7 Subsidence monitoring results at S1

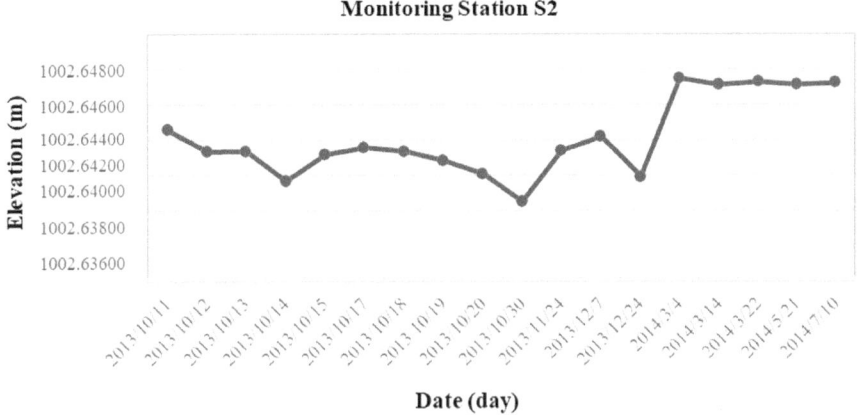

Fig. 4.8 Subsidence monitoring results at S2

4.4.4 Remediation Effectiveness

Five inspection boreholes, approximately 5% of the total number of grouting bore-holes, are completed. The total linear footage is approximately 315 m. Inspection boreholes are constructed within the remediated goaf area. The layout of the inspection boreholes is based on the importance of the buildings, locations with complex geological conditions such as broken strata, collapsed holes, and dropped drill bit; sections with large injection volumes in the last sequence holes, and locations where grouting conditions are abnormal and where analysis shows that there are concerns with grouting quality.

The borehole diameter of inspection boreholes should not be less than 76 mm. The core sampling rate for core run should be greater than 90%. The focus is on judging

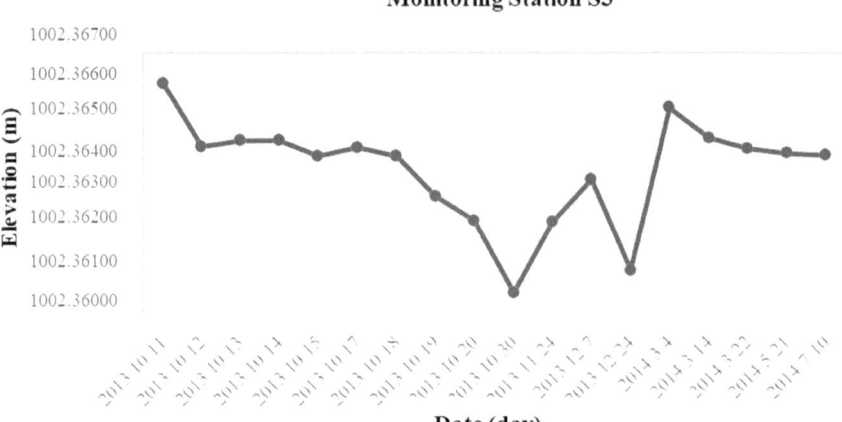

Fig. 4.9 Subsidence monitoring results at S3

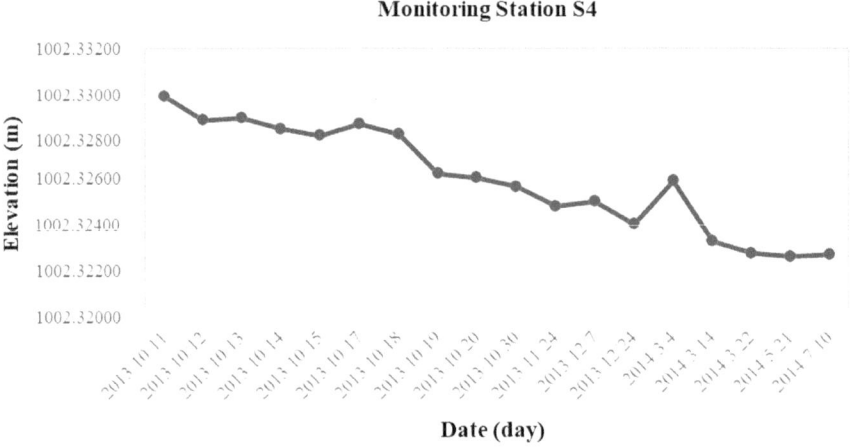

Fig. 4.10 Subsidence monitoring results at S4

whether the drill bit is dropped, whether the circulation fluid is lost, or whether the grouting section is complete. The cores at the goaf layer of each inspection hole are shown in Fig. 4.17.

The slurry filling condition of the goaf is directly observed by coring the entire inspection borehole. The degree of cementation between the slurry stone body and the surrounding rock can be determined based on the drilling core sampling rate and the degree of core fragmentation. The degree of filling of the broken rock layer by the slurry and the integrity after cementation can be determined based on the consumption of circulating fluid during the drilling process, the observation results

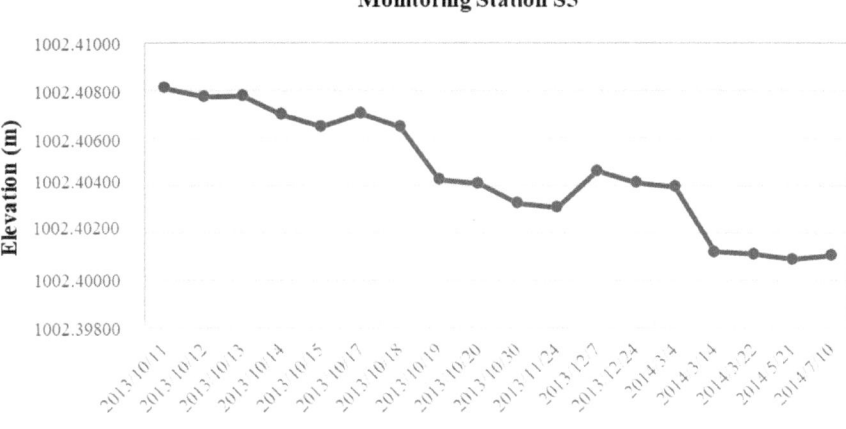

Fig. 4.11 Subsidence monitoring results at S5

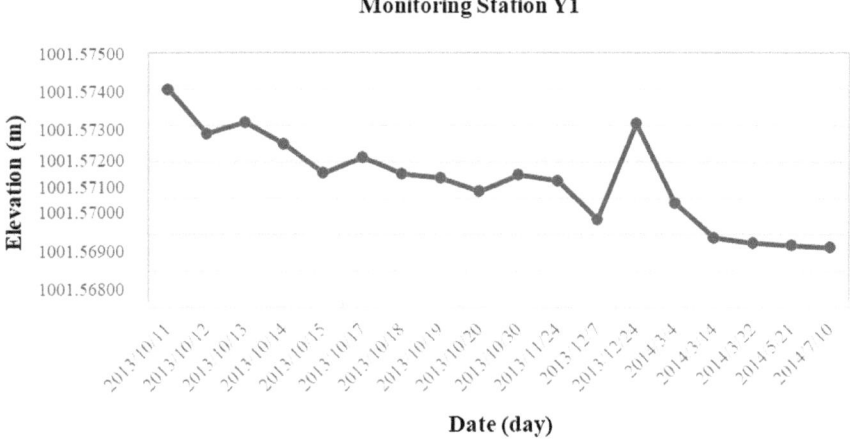

Fig. 4.12 Subsidence monitoring results at Y1

of the static water level in the borehole, and the stability of the borehole wall. The construction conditions of the inspection borehole are shown in Table 4.15.

No drill drops are encountered in the inspection holes. The entire borehole has water return and no water loss. This shows that the underground goaf area has been effectively filled through grouting remediation. The consolidated grout reinforced and supported the strata overlying the goaf area.

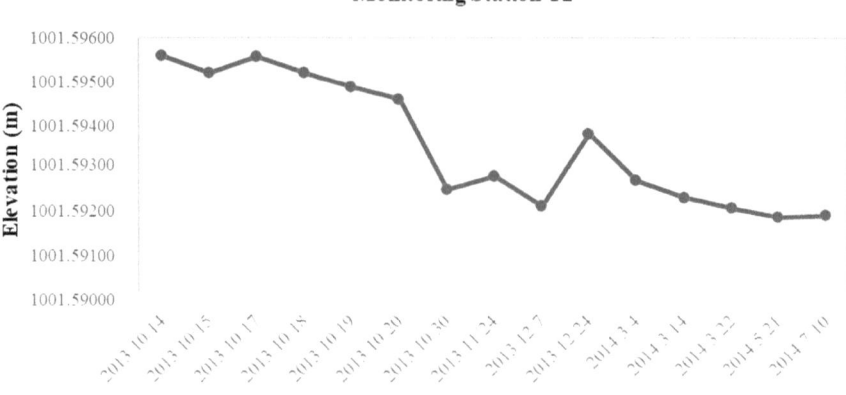

Fig. 4.13 Subsidence monitoring results at Y2

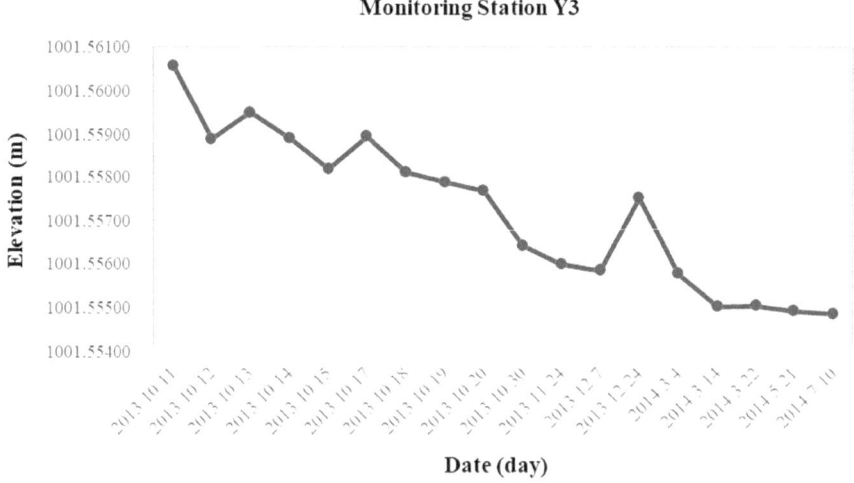

Fig. 4.14 Subsidence monitoring results at Y3

4.5   Conclusion and Recommendations

4.5.1   *Conclusion*

For the goafs left behind by room and pillar mining in most coal mines in northern Shaanxi, the roofs of the goafs fall periodically, and most of the cavities are large. It is difficult for the existing remediation methods and filling materials to achieve the required treatment effects. The complexity and variability of the internal structure of

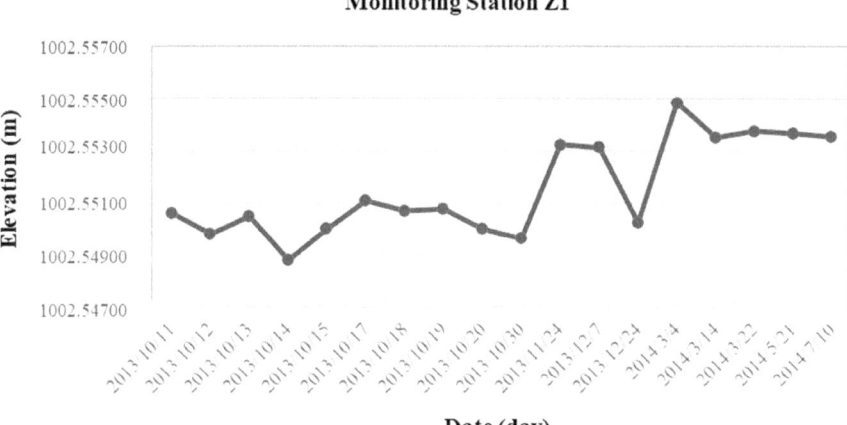

Fig. 4.15 Subsidence monitoring results at Z1

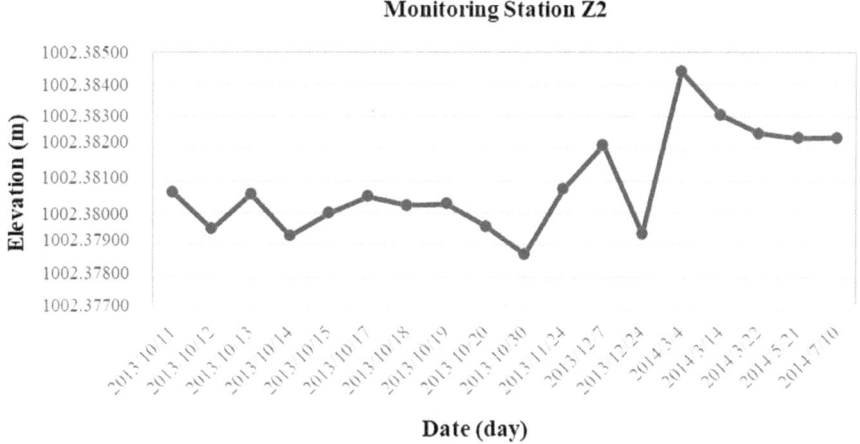

Fig. 4.16 Subsidence monitoring results at Z2

room-and-pillar goafs have posed great challenges to the research and engineering practice of goaf management. The paste materials introduced in this chapter can achieve the effect of filling and treatment, and its remediation effectiveness is verified through the case study. In general, paste materials for goaf remediation have the following features.

- The preparation of paste materials is convenient. The degree of mechanization is high, and the slurry proportion is precise.
- The paste material has low mobility, high consolidation rate, and high controllability of filling range and effect.

Fig. 4.17 Photos of core samples collected from inspection boreholes

Table 4.15 Inspection hole construction

Location	Inspection borehole No.	Peripheral construction borehole No.	Inspection borehole construction	Drill bit drop in neighboring boreholes
Hot air furnace room	Inspection borehole #1	K-17, KZ-26, KZ-25, KZ-23	Borehole depth is 64.2 m, with water return throughout the entire borehole, no water loss, and no drill drop; intercepted paste consolidation body 59–62 m	K-17 borehole: drill bit drop of 4.5 m at 58.5 m KZ-26 borehole: drill bit drop of 3.7 m at 58 m KZ-25 borehole: drill bit drop of 2.8 m at 56.6 m KZ-23 borehole: drill bit drop of 3.5 m at 56 m
	Inspection borehole #2	KZ-27, KZ-28, KZ-24, KZ-25	Borehole depth is 64.8 m, with water return throughout the entire borehole, no water loss, and no drill drop; intercepted paste consolidation body 59–63 m	KZ-27 borehole: drill bit drop of 3.6 m at 56.4 m KZ-24 borehole: drill bit drop of 3.2 m at 56.8 m KZ-25 borehole: drill bit drop of 2.8 m at 56.6 m
Fire pool	Inspection borehole #3	K-9, KZ-21, KZ-16, KZ-19	Borehole depth is 61.5 m, with water return throughout the entire borehole, no water loss, and no drill drop; intercepted paste consolidation body 58–61 m	K-9 borehole: drill bit drop of 4 m at 55 m KZ-21 borehole: drill bit drop of 2 m at 57.7 m KZ-16 borehole: drill bit drop of 4 m at 55 m
Drive room	Inspection borehole #4	KZ-10, KZ-11, KZ-12, KZ-13	The hole depth is 61.5 m, with water return throughout the entire borehole, no water loss, and no drill drop; intercepted paste consolidation body 54.5–57.5 m	KZ-10 borehole: drill bit drop of 1 m at 52.5 m KZ-11 borehole: drill bit drop of 2.4 m at 51.1 m KZ-12 borehole: the hole is 52 m and is drilled 6 m lower

(continued)

Table 4.15 (continued)

Location	Inspection borehole No.	Peripheral construction borehole No.	Inspection borehole construction	Drill bit drop in neighboring boreholes
	Inspection borehole #5	K-1, K-2, KZ-6, KZ-7	The hole depth is 60.5 m, with water return throughout the entire borehole, no water loss, and no drill drop; intercepted paste consolidation body 54.5–57 m	K-1 borehole: drill bit drop of 3.5 m at 54.5 m K-2 borehole: drill bit drop of 3.5 m at 54.5 m KZ-6 borehole: drill bit drop of 1.3 m at 52.2 m KZ-7 borehole: drill bit drop of 0.9 m at 51.5 m

- The paste material is sourced from local materials, and local aeolian sand is selected as the aggregate, which significantly saves material costs.
- The filling process is divided into two types: pressing and pouring, which are convenient to construct and simple to operate.

4.5.2   Recommendations

The paste filling method has been applied to goaf remediation in the study area, achieving good economic and social benefits. The following need to be further studied to improve application of the paste remediation method.

- The relationship between the accumulation morphology, diffusion range and grouting pressure of paste materials.
- Research on filling paste materials with pier-type control technology.

Chapter 5
Subsidence Prevention by Controlling Groundwater in Coal Mines

Abstract Huangsha Mine of north China adopts the multi-level development method with production levels at -280 level and -500 level. The average water flow in the mine is approximately 3.5 m^3/min. A water inrush incident occurred in 2011 at the bottom of the -280 level, and a large volume of groundwater flowed into the mine from the Ordovician limestone underlying the goafs. The peak water flow exceeded 400 m^3/min, which was greater than the drainage capacity. The working face level was completely flooded as well as the pump house at the -500 level. In response to the water inrush, the drainage capacity of the mine increased significantly by installing large-capacity submersible pumps to control the rate of the groundwater level rise. A concurrent grouting program was implemented to fill the goafs and cut off pathways from the water sources to the mining area and the water passageways between Huangsha Mine and the neighboring mines. The water inrush and engineering activities associated with the grouting program caused significant groundwater level fluctuations, which are contributors to collapse pits or subsidence depressions on the ground.

Keywords Long-wall mining · Water inrush · Mine flooding · Grouting · Water passageways

5.1 Introduction

Huangsha Mine is in the southwest of Fengfeng Mining Area of north China. It mainly mines the 2$^\#$ coal seam of Shanxi Formation of Permian System. The mine adopts the multi-level development method of inclined shaft, with current production levels at -280 level and -500 level. The normal water flow rate is approximately 3.5 m^3/min. In December 2011 a working face at the bottom of the -280 level had a water inrush incident. The working face is 850 m long, and the thickness of the coal seam varies from 2.9 to 4.1 m. A large volume of groundwater flowed into the mine from the Ordovician limestone underlying the goafs. The peak water flow exceeded 400 m^3/min, which was greater than the drainage capacity, resulting in a rapid groundwater

S. Dong et al., *Prevention and Reclamation of Mining-Induced Land Subsidence*,
SpringerBriefs in Earth Sciences, https://doi.org/10.1007/978-3-031-78158-2_5

level rise. The working face level was completely flooded as well as the pump house at the − 500 level. In response to the water inrush, the drainage capacity of the mine increased significantly by installing large-capacity submersible pumps to control the rate of the groundwater level rise. A concurrent grouting program was implemented to fill the goafs and cut off pathways from the water sources to the mining area and the water passageways between Huangsha Mine and the neighboring mines. The water inrush and engineering activities associated with the grouting program caused significant groundwater level fluctuations. The water inrush poses great risk to mine safety and operations, while the abandoned goafs and groundwater dynamics can potentially cause collapse pits or subsidence depressions on the ground.

5.2 Geological and Hydrogeological Conditions

5.2.1 Lithology

Within the exploration depth, the exposed strata are as follows from old to new:

- Ordovician: The Ordovician formation consists of three sections. The first section (O_12_f) is a set of brown-yellow and light red dolomite limestone and dolomite brecciated limestone. The brecciated limestone is mainly dolomite, followed by limestone, with clear edges and corners, different sizes, and tight cementation. After weathering, honeycomb dissolution is developed, with a layer of 2–4 m pure limestone in the middle and a layer of 2–3 m unstable dolomite limestone at the bottom. The thickness varies from 40 to 72 m with an average of 55 m. The second section (O_22_f) is composed of strata that are interbedded with dark gray medium-thick pure limestone and spotted limestone. The spots gradually become smaller from bottom to top, from brown-yellow and gray, yellow to gray-white and white. The pure limestone smells like rotten eggs when hammered. The top and bottom pure limestones often contain flint nodules, which are scattered on the layers. The thickness varies from 63 to 95 m with an average of 85 m. The third upper section (O_32_f) is composed of striated limestone with clear rhythm and light yellow or white. The lower part is brecciated limestone with clear breccia after weathering and dolomite cement. The thickness varies from 8 to 28 m with an average of 18 m.
- Carboniferous: The carboniferous formation consists of the Middle Carboniferous Benxi Formation (C2b) and Upper Carboniferous Taiyuan Formation (C3t). The Benxi Formation is composed of light gray to dark gray siltstone, containing a layer of terminal coal, with a thickness of 0–0.11 m. The middle and lower parts are gray bauxite with oolitic structure, and the surface shows rust-colored oxide film after weathering. The bottom is unstable Shanxi-type iron ore, which is present in the form of nodules or lenses and is sometimes replaced by purple-red iron-bearing mudstone. The thickness of this section ranges from 5.79 to 20 m, with an average of 9 m. The Taiyuan Formation was deposited on the Benxi

Formation, with thickness between 111.77 and 132.57 m and an average of 124 m. This section is the main coal-bearing stratum. The Taiyuan Formation is a series of black-gray siltstone, gray-white medium-fine sandstone interlayers, and 6–8 layers of marine thin limestone in the middle. It is a marine-continental interphase deposition, with a total of 12 coal layers. The mineable coal seams and limestone are deposited stably and are marker layers.

- Permian: The Permian formation includes Shanxi Formation (P1s), Shihezi Formation (P1x), Shiqianfeng Formation (P2sh). The Shanxi Formation is another coal-bearing stratum with 2–5 coal seams. The lithology includes siltstone and medium-fine-grained sandstone sediments. The Shihezi Formation is a set of yellow-brown, purple-red siltstone, intercalated with gray-green medium-fine sandstone and purple red oolitic bauxite, with an average thickness of 54 m, ranging from 34.22 to 64.3 m. Shiqianfeng Formation is composed of purple-red siltstone, interbedded mudstone and thin fine sandstone. The fine sandstone is mainly composed of quartz, feldspar and mica, with well-developed horizontal stratification and oblique stratification. The siltstone and mudstone contain calcareous nodules. The middle part is interbedded with 3–12 m thick layers of marl and mudstone. The marl is generally 2–3 layers, the color is grayish white, with purple spots in some parts, and sometimes contains limestone breccia. The lower part is composed of dark purple mudstone and fine-grained sandstone. The mudstone is rich in ginger-shaped calcareous nodules. The bottom is a layer of dark purple or grayish yellow medium-grained sandstone, which often contains flint and a small amount of mudstone and sandy mudstone fragments. It is calcareous cemented and has an average thickness of 126 m.
- Tertiary: The Tertiary Formation is unconformably above the Permian strata. The thickness varies from 0.20 to 160 m with an average of 75 m. The lithology is composed of conglomerate, clay, sub-clay, sub-sand, sand, fine-medium-coarse sand, and the sand layer is semi-cemented, changing with the terrain, and the thickness gradually increases from west to east. The bottom is quartz sandstone, conglomerate or claystone.
- Quaternary strata (Q): The Quaternary deposit is unconformably above the Tertiary System and is mainly composed of loose sediments such as gray-yellow and red sand, clay, and loess. It is mainly distributed in river valleys, with thin layers on mountain tops and slopes. The thickness varies from 0 to 43 m with a typical value of 5 m.

5.2.2 Geologic Structures

The geologic structures include faults, folds, and collapsed columns. Faults in the Huangsha mining area are well developed, with 50 faults having displacements of greater than 10 m. Tunnel construction and coal mining revealed more than 228 small faults. They are tensile high-angle normal faults with strikes from N 0° to 25° E. The dip angle is generally between 65 and 80°. Small faults tend to be developed between the large faults. Three folds were identified, and they are North Huangsha syncline, Nanhuangsha anticline, and Majiahuang syncline.

- North Huangsha syncline is in the Huangsha mining area with the syncline axis on the north side of the main inclined shaft. It has been confirmed by drilling and − 100 level mining tunnels that the folds are gentle with inclination angles between 13 and 19° on both wings.
- Nanhuangsha anticline is in the southern part of Huangsha Mine and north of two villages. The axis roughly turns from SW to N 60° E, with a dip angle of approximately 15°.
- Majiahuang syncline has a trike of EW and strata dip angles between 13 and 21° on both wings. It is verified by exploratory holes and three-dimensional seismic data.

Collapse columns are unique geologic features in the area. Three collapse columns are reported, and they are non-water-conducting and non-water-containing. The largest one has an equivalent diameter of 80 m and a major-minor axis ratio of 2:1. The other two are 150 m apart. The diameter of both columns is approximately 35 m.

5.2.3 Hydrogeology

Both aquifer and aquitards are important in mine-induced water inrush and subsidence investigation and remediation. Table 5.1 summarizes their characteristics.

The water-resistance capability of the aquitards weakens after coal seam mining, especially in caved zone and fractured zone, which are often referred to as water-conducting fracture zone. The presence of large and medium-sized water-conducting zones and water-conducting collapse columns tend to destroy the water-blocking capacity of aquitards. The relationship between the aquiclude combinations is shown in Fig. 5.1.

5.3 Design of Sealing Groundwater Passageways

5.3.1 Analysis of Water Inrush

Source of water and passageways are two critical elements in water inrush analysis. The source of the water inrush is from the Ordovician limestone aquifer based on the following:

- The water inrush is dramatic with the maximum flow rate at 400 m^3/min. The large water volume lasts for a long time, causing water level to rise to − 45 m in the mine.

Table 5.1 Characteristics of main aquifers in Huangsha Mine

Aquifer or aquitard	Lithology and characteristics
XII aquifer	Quaternary loose mud layer sedimentary aquifer with thickness between 2 and 5 m. Based on pumping tests, the maximum water yield can reach 19 L/s m. Hydraulic conductivity varies from 35 to 202 m/d
Aquitard 1	It is composed of clay, sub-clay and loess, with a thickness of 40 ~ 75 m and good water-barrier performance
XI aquifer	Tertiary fracture aquifer with thickness between 0.2 and 160 m and water yield of 0.375 L/s m
Aquitard 2	It is composed of medium-fine grained sandstone, siltstone and mudstone, with a thickness of 211 ~ 242 m and good water isolation
X aquifer	Sandstone fissure aquifer at the bottom of the third member of the Upper Shihezi Formation. Water yield is approximately 0.05 L/s m, and hydraulic conductivity is 0.1 m/d
Aquitard 3	It is composed of siltstone and mudstone with a thickness of 15 ~ 30 m and good water-barrier performance
IX aquifer	Sandstone fissure aquifer in the middle of the second section of the Upper Shihezi Formation. Water yield ranges from 0.018 to 0.14 L/s m. Hydraulic conductivity varies from 0.1 to 0.7 m/d
Aquitard 4	It is composed of medium-fine sandstone, siltstone and mudstone, etc. It is 126 ~ 174 m thick and has good water-barrier performance
VIII aquifer	Sandstone fissure aquifer at the bottom of the Upper Shihezi Formation. The thickness varies from 8 to 15 m. Aquifer is confined, and the water yield is poor
Aquitard 5	It is composed of siltstone, bauxite siltstone and mudstone. It is 29 ~ 41 m thick and has good water insulation
VII aquifer	Sandstone fissure aquifer at the bottom of the Lower Shihezi Formation. Thickness ranges from 2 to 16 m. Water yield is greater than 0.017 L/s m. Hydraulic conductivity varies from 0.2 to 103 m/d
Aquitard 6	It is composed of siltstone, medium-fine sandstone and coal seams, etc. It is 10 ~ 19 m thick and has poor water insulation
VI aquifer	Sandstone fissure aquifer in the roof of coal seam #2. Thickness ranges from 10 to 20 m. Water yield varies from 0.008 to 0.36 L/s m. Hydraulic conductivity varies from 0.04 to 3.97 m/d
Coal seam #2	Coal seam thickness 2.25 ~ 6.15 m, average 4.01 m
Aquitard 7	Mainly composed of siltstone, mudstone and fine-grained sandstone. Thickness varies from 22.71 to 56.13 m, good water insulation
V aquifer	Taiyuan Formation Yeqing limestone fracture aquifer. Thickness ranges from 1.48 to 4.9 m, average of 3 m. Maximum water inflow into mine is approximately 120 m^3/h
Aquitard 8	It is composed of mudstone, siltstone, fine-grained sandstone and coal seams. It is 26 ~ 42 m thick and has good water insulation
IV aquifer	Taiyuan Formation mountain and Fuqing limestone fracture aquifer. Thickness varies from 1 to 2 m. It is strong aquifer with hydraulic conductivity ranging from 0.46 to 0.53 m/d

(continued)

Table 5.1 (continued)

Aquifer or aquitard	Lithology and characteristics
Coal seam #6	Thickness 0.59–1.18 m, average thickness 0.89 m
Aquitard 9	Siltstone, mudstone and medium-fine sandstone, with a thickness of 9.28–18.67 m and good water-proof performance
III aquifer	Taiyuan Formation Xiaoqing limestone aquifer. Thickness is 1 m. Water yield is poor
Aquitard 10	It is composed of siltstone, fine-grained sandstone and coal seams, etc. It is 18 ~ 26 m thick and has poor water-proof performance
II aquifer	Daqing limestone karst fissure aquifer. Thickness varies from 3.21 to 8.07 m. Water yield ranged from 0.7 to 0.9 L/s m. Hydraulic conductivity varies from 1.6 to 1.7 m/d
Aquitard 11	It is composed of bauxite, siltstone and coal seams. The thickness is 14 ~ 28 m, and the water-proof performance is poor
I aquifer	Ordovician limestone karst fissure aquifer. This aquifer is a confined aquifer at the base of the coal-bearing strata. The average thickness is 500 m. The karst caves and fissures are more extensive in the upper section with water-richness decreasing with depth. Hydraulic conductivity is 68.61 m/d. Water yield is 105 L/s m

- The groundwater level at an Ordovician limestone monitoring well, approximately 900 m away from the water inrush point, decreased by 0.5 m 16 h after the water inrush. The total water level drop is 10 m.

No water-conducting structures are found during the mining process of the working face. The water inrush was a delayed water inrush in the goafs after completion of the working face. The water inrush passageway is a concealed water-conducting structure in the coal floor. The nature and specific location of the water-conducting structure are to be verified during the grouting of the water inrush channel.

5.3.2 Strategy of Sealing Groundwater Passageways

Immediately after this major water inrush occurred, engineering measures are taken to minimize possibility that the entire mine is flooded and water flows into surrounding mines. An emergency response was conducted to increase dewatering capability to slow down the rising rate of the water level and control the rise of the water level, to cut off the water passageways between the Huangsha Mine and the adjacent mines to ensure the safety of neighboring mines, to implement water plugging projects to solve the water inrush problem.

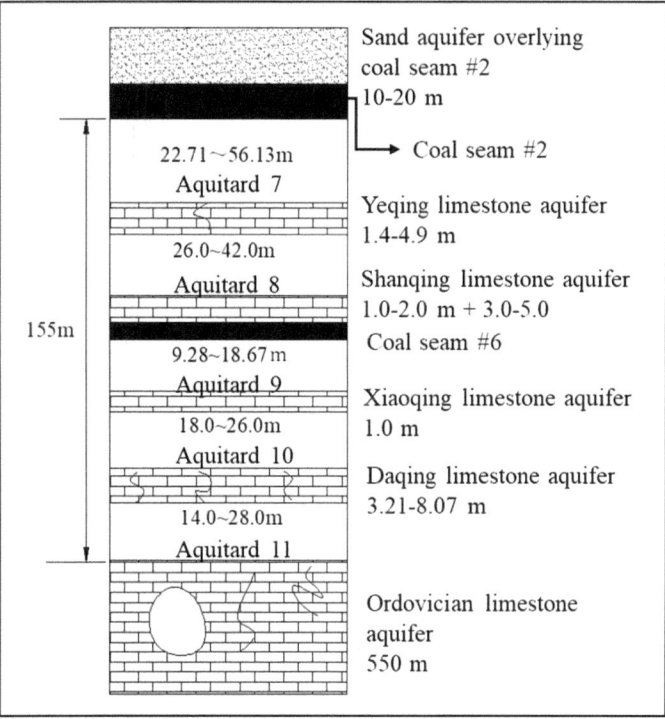

Fig. 5.1 Relationships between major aquifers and aquicludes

Although the nature and location of the water inrush passageways are unknown, speedy implementation of the water plugging project is demanded to reduce dewatering pressure and costs. The overall strategy is as follows:

- First, intercept the flow in the mine tunnels, inject aggregate in the material transport tunnel and chute to form a water-blocking wall, and then reinforce it through grouting to block the path from the Ordovician limestone to the mine through tunnels.
- Seal the tunnels again and conduct grouting and sealing work on the water bursting passageways in the formation underlying the working face to mitigate the water inrush.

The difficulty of the water plugging project is the large amount of water from the Ordovician limestone. According to the drainage capacity of the mine, the water inrush from the Ordovician limestone is approximately 180 m^3/min at the dynamic equilibrium water level of − 48 m. The water plugging project needs to be implemented in two tunnels, the material transport tunnel and the chute. The potentiometric pressure of the Ordovician limestone aquifer at the water plugging locations is approximately 1.7 MPa.

5.3.3 Tunnel Interception Design

Eight boreholes are arranged to intercept the material transport tunnel and chute with four boreholes for each target. The borehole diameter is φ311 mm, and a φ244.5 × 8.94 mm orifice pipe is inserted 5 m into the bedrock to isolate the Quaternary strata. In the bedrock section at depths between 170 and 740 m, the borehole diameter is φ216 mm with φ178 × 8.05 mm through-hole casing. The boreholes are advanced to approximately 50 m above the tunnel roof, the casing is installed. In the open borehole section from 740 to 790 m the borehole diameter changes to φ155 mm. A drilling bit of φ152 mm is advanced through the tunnel. The section between 744 and 754 m is reinforced using the φ152 mm drilling bit. All boreholes must be drilled from the ground and accurately penetrate the target tunnel, otherwise the boreholes are useless. The borehole inclination check should be conducted regularly. If excessive deviation is found, effective measures should be taken to make corrections to create conditions for aggregate injection and grouting. The casings used in the project are double layered. The first layer of casing isolates the alluvial layer and is required to be installed into the bedrock for no less than 5 m. The second layer of casing is required to be installed 50 m above the tunnel roof. All levels of casing are constructed with positive circulation cement slurry. The intercepting bore structure is detailed in Fig. 5.2.

Preliminary hydrogeological observations are required during drilling. If a large amount of water leakage, drill bit drop, drill bit burial, air suction, or water gushing occur, their depth, layer, and water consumption are recorded in detail. After drilling through the tunnel, the height of the tunnel is measured with the drilling tool. All ground boreholes are flushed with water for no less than 30 min after completion. The static water level of the borehole is measured and recorded.

5.3.4 Technical Approach of Grouting

5.3.4.1 Aggregate Feeding Design

The grouting engineering includes aggregate injection and slurry injection. The aggregates include sand, fine-grained stone, crushed stone. Crushed stone is divided into four types: 0.5, 1–2, 1–3, and 2–4 cm. Before pouring aggregate, clean water should be poured for at least 15 min. Flushing must also be conducted for 15 min after feeding the aggregates. When feeding aggregates, the water supply shall not be less than 120 m³/h. When feeding the aggregates, a trial test should be conducted for optimization of particle sizes. The fine aggregate should be injected first. The aggregate injection generally requires a uniform injection speed. If necessary, intermittent quantitative injection is performed. The aggregate injection stops after 1 h of injection, and then water is flushed into the borehole.

Fig. 5.2 Construction specifications of intercepting boreholes

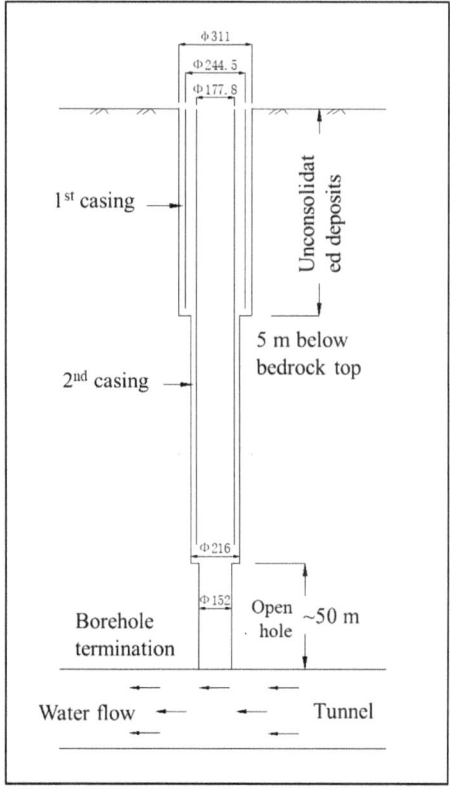

During the aggregate injection period, the amount of aggregate injected should be counted and the possible tunnel filling section should be estimated to guide the next aggregate injection plan. The water–solid ratio should be maintained. Injection without water is strictly prohibited. Pouring a certain type of aggregate should avoid mixing it with large-sized aggregate and surface mud. On-site construction personnel closely observe the drilling feed situation, stop injection immediately if any abnormal situation occurs, and then flush the hole with water. In case the borehole is blocked by aggregate, borehole cleaning should be conducted immediately to open the blockage.

Silicate PO42.5 cement is used in cement slurry injection. When the aggregate is fully filled in the tunnel, it is expected that the aggregate at the bottom of the hole has reached the top. The mine water inrush in the tunnel changes from turbulent flow to laminar flow. Pouring of cement slurry starts when the water inrush volume is reduced to approximately 200 m^3/h. During grouting, the slurry is thin at first and then thick. The water-cement ratio of the slurry is adjusted in time according to the amount of slurry consumed. The overall water: cement ratio of the slurry is in the range of 2:0.7. Depending on the slurry injection conditions, intermittent grouting method can be adopted. Grouting ends when the grouting is 3 MPa and the pump volume ranges from 40 to 60 L/min for 30 min.

5.3.4.2 Grouting Design of Water Inrush Passageways

Eight boreholes are designed to investigate the groundwater passageways for the inrush. The specific locations of the drilling holes are to be determined based on the progress of the project. The boreholes are advanced with following specifications:

- Loose layer section: The borehole diameter is 311 mm. A 244.5 × 8.94 mm orifice pipe is inserted into the bedrock section, entering 5 m into the bedrock to isolate the Quaternary strata.
- Bedrock section: The borehole diameter changes to φ216 mm. Install 178 × 8.05 mm through-hole casing and consolidate the second layer of casing to the hard rock layer about 20–50 m above the water-conducting structure according to the drilling exposure during drilling.
- Open hole section: The borehole diameter is φ155 mm. A drill bit of φ152 mm is used to advance to the bottom of the borehole.

Directional drilling is required during the drilling process. The target area should be intercepted as required. If there are branch boreholes, the branch boreholes must hit the target area according to the entry angle required by the technology. The specific requirements are to be determined based on the drilling exposure. The borehole is generally double-cased. The first layer of casing isolates the alluvial layer and is required to be at least 5 m deep into the bedrock. The second layer of casing is required to be placed in a hard formation about 20–50 m above the water-conducting fractures. Each level of casing is required to be positively circulated and installed with cement slurry. After the casing is lowered, a 152 mm drill bit is used to continue drilling. If a water-conducting structure such as a collapse column is detected, a drill bit of φ216 mm diameter should be used to expand the borehole with to a hard rock layer 20–50 m above the top of the structure.

Preliminary hydrogeological observations are performed during drilling. If large amounts of water leakage, drill bit drop, drill bit burial, air suction, or water gushing occur, their depth, layer, and water consumption are recorded.

5.3.4.3 Adaptive Management Approach

The grouting and water plugging effort is an emergency rescue project. Due to the urgency nature and the complexity of hydrogeological conditions, the variability of water flow in the tunnels during diversion and the particularity of the grouting project, design parameters such as the number of boreholes, borehole locations, drilling technology, and grouting technology need to be adaptive to new data available. Design changes are inevitable. The design parameters are adjusted in time during the implementation of the project according to the exposure of the tunnels and strata during the progress of the project and the water blocking effect.

During the implementation of the tunnel interception project, due to the large amount of water and high potentiometric pressure of the water inrush, adding

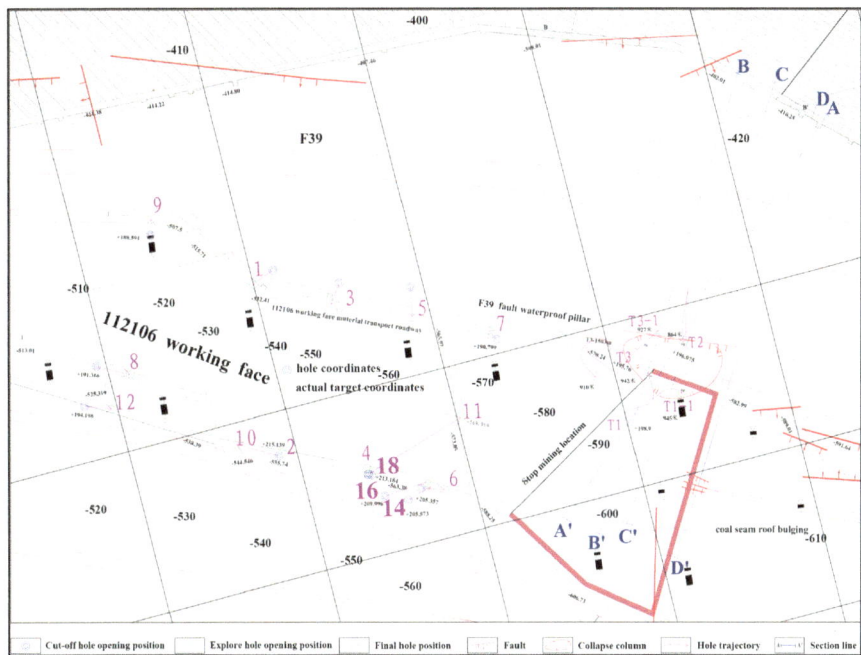

Fig. 5.3 Borehole layout in investigation and remediation

four boreholes (9, 10, 11 and 12) were added to increase the length of the tunnel interception section.

The cross-sectional size of the tunnel through which the water burst occurred was larger than anticipated. With a tunnel width of 4.5 m, it was difficult to connect the aggregates at the end of the tunnel interception and plug the passageways. Three more boreholes (14, 16 and 18) were added between boreholes 4 and 6 in the lower tunnel. Figure 5.3 shows the final borehole layout.

Table 5.2 summarizes the drilling results. The tunnel interception is the first step of the water mitigation effort. A total of 15 boreholes were constructed to intercept the tunnels. The total drilling footage is approximately 11,619 m. Because of the urgency of plugging the tunnels for mine safety, some of the added boreholes were drilled directly to the tunnel without inserting a second casing.

5.3.4.4 Challenges in Tunnel Interception and Water Plugging

There is a significant difference between the drilling to intercept tunnels and the drilling in typical geological explorations. In geological exploration, the drilling rig is usually mobilized after the borehole drilling is completed, and additional drilling and borehole sweeping are not required. However, when the interception boreholes are drilled, aggregates need to be injected through them. Blockage often occurs

Table 5.2 Summary of boreholes intercepting tunnels

Borehole No.	Casing depth (m)		Borehole depth (m)	Notes
	1st casing	2nd casing		
1	162.97	674.0	734.0	• Loss of water: 718 m • Intercepting tunnel: 719.8–724 m • Drill bit drop: 4.2 m
2	173.22	701.0	782.0	• Loss of water: 765.9 m • Intercepting tunnel: 769.2–772 m • Drill bit drop: 2.8 m
3	163.3	684.6	742.1	• Intercepting tunnel: 733–737 m • Drill bit drop: 4.1 m
4	171.22	709.83	780.5	• Loss of water: 774 m • Intercepting tunnel: 776.81–780.56 m • Drill bit drop: 3.75 m
5	159.34	713.3	794.0	• Intercepting tunnel: 749–753 m • Drill bit drop: 4 m
6	No casing	751.4	780.8	• Intercepting tunnel: 777–780.8 m • Drill bit drop: 3.8 m
7	168.16	704.45	871.4	• Loss of water: 757.6 m • Intercepting tunnel: 759.9–764.4 m • Drill bit drop: 4.5 m
8	No casing	688.83	728.9	• Intercepting tunnel: 715–718.85 m • Aggregate at bottom of 1.2 m • Drill bit drop: 3.85 m
9	151.4	641.13	705.1	• Intercepting tunnel: 691–695.1 m • Drill bit drop: 4.1 m
10	172.52	702.95	777.3	• Loss of water: 761.5 m • Intercepting tunnel: 764.3–767.3 m • Drill bit drop: 3.0 m
11	No casing	703.38	801.5	• Loss of water: 787 m • Intercepting tunnel: 787–791.5 m • Drill bit drop: 4.5 m

(continued)

Table 5.2 (continued)

Borehole No.	Casing depth (m)		Borehole depth (m)	Notes
	1st casing	2nd casing		
12	156.7	668.0	736.8	• Encountered foreign objects at 721 m • Intercepting tunnel: 722.8–726.8 m • Drill bit drop: 4 m
14	160.5	763.46	793.7	• Intercepting tunnel: 779.0–783.7 m • Drill bit drop: 4.7 m
16	168.05	768.11	797.8	• Intercepting tunnel: 783.4 m • Drill bit drop: 4.4 m
18	No casing	763.68	792.6	• Intercepting tunnel: 779.5–782.6 m • Drill bit drop: 3.1 m

during the aggregate injection, especially in the later stage when the volume of the tunnel becomes smaller, blockage is particularly prone to occur before the aggregates in the tunnel are connected. For example, borehole No. 2 was swept 19 times, while borehole No. 10 was swept 17 times. Therefore, the drilling rig is required to cooperate with the grouting to repeatedly sweep the hole until the aggregates in the borehole are broken and swept away. After the borehole is drilled to the bottom again, the drill tool is again raised to standby, or the tools are mobilized to sweep other boreholes.

The borehole sweeping is different from conventional drilling. It not only involves a extensive labor work, but also has high requirements for drilling technology. Drilling misshapes such as drill bit jamming, borehole collapse, and drill bit drop are prone to occur during the borehole sweeping process. Once these drilling incidents occur, not only will the drilling rig be potentially damaged, but the drilling tools may also fall into the boreholes. Consequently, new boreholes or branch boreholes need to be drilled, resulting in delays in the rescue and water blocking timeline. Other challenges include the following:

- Drill bit jam: Drill bit jam occurred in several boreholes during the drilling process. Because most of the drilling rigs in this emergency rescue project are powerful with the lifting force, the drill tools were eventually forcibly pulled out without causing any losses.
- Borehole collapse: Borehole collapse occurred in many boreholes including No. 5, No. 2, No. 10, No. 4, and No. 12 during the sweeping process. Once the borehole collapses, the bare interval becomes empty. The drilling mud that carries rock cuttings is unable to be taken out of the borehole in normal operations. The drilling rate is extremely slow. Where the borehole collapse occurs, the correction

measures are to grout and reinforce the borehole. The borehole is then re-swept after the strength is increased.

- Drill drop: Drill drop occurred in boreholes No. 8 and No. 12 during sweeping process. The drill bit dropped twice in borehole No. 8. Because the drilling rods and tools are at great depths when the drill bit dropped in borehole No. 12, the drilling tools could not be retrieved. A branch borehole is drilled at a depth of 682 m.

After the aggregates are connected in the tunnels, water flow in the tunnels becomes limited. The permeability of the tunnels decreases significantly. As a second step of the remediation effort, cement slurry is used for grouting reinforcement immediately. The reinforced sections are mainly the voids between aggregates in the tunnels, the gaps between aggregate sections and the tunnels, and the weak zones and water-passing sections immediately around the tunnels.

Reinforcement grouting drilling requires drilling to 10 m below the tunnel floor. If there is water loss during drilling, grouting will be carried out immediately. Then the slurry will be drilled through and down into the tunnel aggregates. If there is water loss in the tunnel section, grouting will be conducted again to reinforce the borehole. Then drilling continues to 10 m below the tunnel floor. If there is no water loss, drilling goes directly to 10 m below the tunnel floor and grout to meet the termination criteria.

The hole sweeping phenomenon also exists in the reinforcement grouting drilling, Except for sweeping away aggregates when passing through the tunnels, the rest of the borehole sweeping sections are drilled into the cement slurry filling. As a result, the borehole sweeping is not as challenging, and the number of hole sweeping is also fewer.

5.3.4.5 Aggregate Feeding Process

Aggregate injection is a key step in tunnel interception. Under dynamic water conditions, the premise for achieving effective grouting is to condition the tunnels to reduce the water flow rate and thus erosion to cement slurry. Otherwise, any injected cement slurry would be carried to long distances by high water flow, resulting in ineffective dispersion. The solution is to inject various sizes of aggregate into the tunnel through boreholes. The type of aggregate includes coarse and fine sand, slag of various particle sizes, broken bricks, and others. The purpose is to fill and plug the channels to reduce permeability of the tunnels. The flow dynamics in the tunnels change from turbulent to laminar to facilitate the effective accumulation and consolidation of the injected slurry. The aggregate injection process is shown in Fig. 5.4.

The aggregate injection starts with pumping water into the borehole for 15 min to confirm that the pipeline and the borehole are unobstructed. Both water and aggregates are injected simultaneously with the water: solid ratio between 6:1 and 10:1. The amount of aggregate filling must be strictly controlled during feeding. The aggregate feeding rate must be uniform and match the water flow rate. It is prohibited to

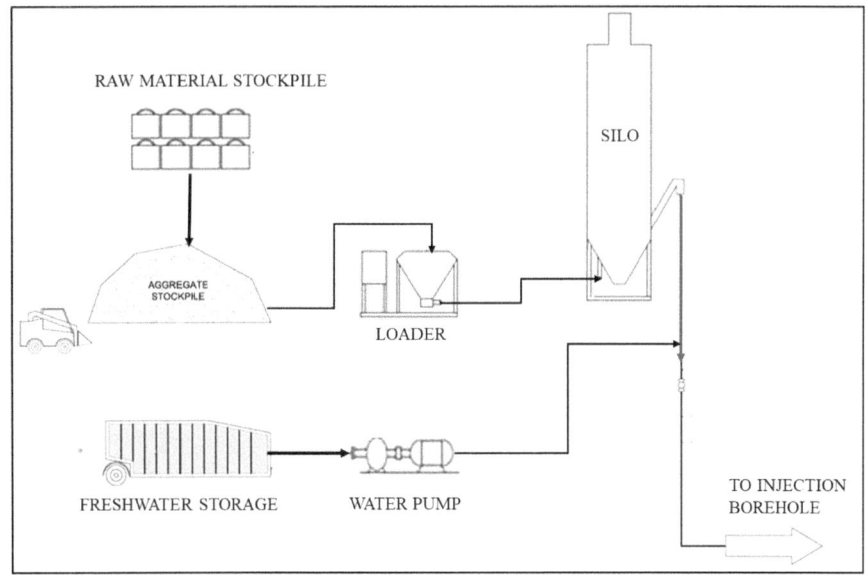

Fig. 5.4 Schematic diagram of aggregate pouring process

pour aggregate too quickly or in excess to prevent clogging of the borehole. During the feeding process, if the amount of water injected into the funnel is reduced or the feeding speed is significantly slowed down, the pouring should be stopped immediately to check whether the funnel or pipeline is blocked by large-diameter aggregate. Aggregates must be sorted before injecting to ensure that the aggregate particle size meets the requirements. Table 5.3 summarizes the quantities of aggregates in six boreholes.

As the cross-section of the tunnels becomes smaller with aggregate injection, the hydraulic gradient increases sharply, which causes the aggregate walls to breach. Table 5.4 summarizes four breach events in which the tunnel failed to be completely

Table 5.3 Summary of aggregate feeding volume (m^3) in boreholes

Borehole No.	Sand	Fine stone	#5 Stone	#1 and 2 Stones	#1–3 Stones	Stone powder	Total
2	1818	3485	740	15	–	–	6058
4	846	3539	7210	–	–	230	11,825
6	430	840	320	–	–	–	1590
8	420	1050	1200	2700	–	–	5370
10	110	230	1730	3010	100	–	5180
12	1907	947	–	–	–	–	2854
Total	5531	10,091	11,200	5725	100	230	32,877

Table 5.4 Summary of plugging breach in tunnel interception

Breached time	Before water-plug is established in tunnel				After breach of the water-plug section in tunnel	
	Water level in borehole (m)	Water level in mine (m)	Water level in Ordovician limestone (m)	Flow rate in tunnel (m^3/min)	Water level in borehole (m)	Flow rate in tunnel (m^3/min)
March 6	+ 46.5	− 40.18	+ 116.14	75.83	+ 14.2	90.8
March 14	+ 50.2	− 39.18	+ 116.10	83.3	+ 1.9	89.17
March 20	+ 105.6	− 37.38	+ 116.47	43.8	+ 76.6	74.1
March 23	+ 106.7	− 37.59	+ 116.52	52.83	+ 43.8	71.17

plugged. A breach of water-plug in the tunnel corresponds with a water level drop in the borehole and an increase of flow rate in the tunnel. The water-plug is established after boreholes #14, #16 and #18 are drilled and used for aggregate injection. The flow rate reduced to approximately 26 m^3/min, and the water levels in boreholes #7 and #10 rose to + 121.3 m and 119.74 m, respectively, exceeding the water level in the Ordovician Limestone of + 116.67 m.

5.3.4.6 Cement Slurry Grouting Process

The cement slurry grouting adopts static pressure grouting method with borehole inlet sealed. Once the aggregates are judged to be connected to the tunnel top, the borehole is immediately closed, and cement slurry is injected from the borehole inlet. During grouting, the water: cement ratio of the slurry is adjusted. Thin slurry is injected first, followed by thick slurry. The grouting stops immediately after the pressure is applied to the top of the tunnel. The slurry is allowed to mature. Then the borehole advances into the tunnel or 10 m below the tunnel floor, and then performs grouting again. If pressure builds up at the orifice during grouting the formation 10 m below the tunnel floor, grouting terminates after final pressure reaches 3 MPa and injection rate reduces to 40–60 L/min, and these number maintain for 30 min. At completion of the slurry grouting, approximately 28,255 t of cement is injected into the tunnels to reinforce the water-plug.

5.4 Investigation and Grouting of Water Inrush Passageway

The water inrush passageway is unknown and needs to be located prior to remediation. The investigation of the passageway mainly relies on the drilling method and the comprehensive analysis of the strata that are exposed by drilling. The exploration boreholes are also used as grouting holes to seal the water inrush passageway. Cement slurry and cement-fly ash mixed slurry are used for grouting. Drilling and grouting are conducted concurrently in this project.

5.4.1 Investigation of Water Inrush Passageway

Three boreholes, labeled as T1, T2, and T3, were drilled to investigate water inrush passageway. Boreholes T1 and T3 have a branch borehole each, labeled T1-1 and T3-1, respectively. The locations of the boreholes are shown in Fig. 5.2. The boreholes are specifically constructed, as shown in Fig. 5.5, for water inrush passageway investigation.

- Loose layer section: The borehole diameter is φ311 mm. A φ244.5 × 8.94 mm orifice pipe is inserted into the bedrock section, entering 5 m into the bedrock to isolate the Quaternary deposits.
- Bedrock section: The borehole diameter is φ216 mm with φ178 × 8.05 mm casing. The second casing is installed approximately 20–50 m above the water-conducting fractured zone.
- Open section to borehole bottom: The borehole diameter is φ155 mm with a drill bit of φ152 mm.

Due to the existence of branch boreholes, secondary casing was not lowered in T1-1 borehole and T3-1 borehole during construction. The secondary casing of T1 borehole was installed to a depth of 787.64 m, whereas the secondary casing of T2 borehole was installed to a depth of 809.61 m. The secondary casing of T3-1 hole was lowered to a depth of 791.6 m.

T1 borehole: The opening position of this borehole is located above the goaf of the working face. It is the only borehole that passes through the goaf and drills into the floor formation underlying the working face. Because this borehole passes through the goaf, it has undergone multiple grouting and borehole sweeping in the overburden overlying the coal seam before passing through the goaf and entering the coal seam floor formation. A major water circulation loss occurred at 728 m. The drill bit dropped by 1 m between 755 and 756 m. The rock strata between 777 and 785 m were broken, which is interpreted to be the coal seam goaf section. The rock strata after 785 m were intact. A coal layer with a thickness of 0.5 m was found at 819 m. Yeqing limestone was found between 824 and 827 m. A coal seam was

Fig. 5.5 Construction specification of boreholes for water inrush passageway investigation

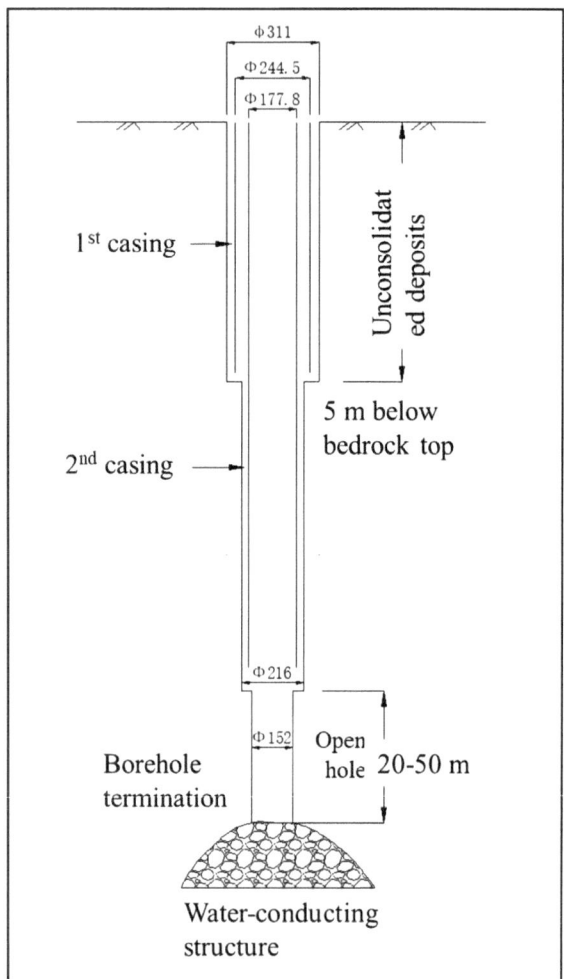

encountered at 872 m with a thickness of approximately 1 m. Fuqing limestone was found between 874 and 878 m. Xiaoqing limestone was found from 891 to 892.5 m. A major water loss occurred at 913 m. The Ordovician limestone was encountered at 929 m. The final depth of this borehole is 942 m.

T2 borehole: This borehole is arranged outside the working face. The drill bit dropped 4 m in the borehole between 844 and 848 m. Bauxite mudstone was found between 860.5 and 862 m. The Ordovician limestone was encountered at 862 m. The borehole was swept several times during grouting. The terminal borehole depth is 864 m.

T3 borehole: The borehole is drilled into the coal body in front of the working face, near the tunnel and goaf. Fuqing limestone was found at intervals of 870 and 874 m. No major water loss is encountered until 910 m where grouting started. The

Ordovician limestone was not intercepted. During the grouting, grouting slurry was found in the goaf of T1 borehole. Then the borehole was sealed, and branch boreholes were drilled. The final depth of this borehole is 910 m.

Borehole T1-1: Borehole T1-1 is a branch borehole of T1. The branch starts at 803 m. A layer of coal was found at 819 m with a thickness of about 0.5 m. Yeqing limestone was encountered between 823 and 827 m, while Fuqing limestone was found between 873 and 877 m. Xiaoqing limestone was found at intervals of 890 and 892 m. Daqing limestone was encountered between 910 and 915 m. The Ordovician limestone was intercepted at 930 m. A major water loss occurred at 942.5 m. The borehole is terminated at 945 m.

Borehole T3-1: Borehole T3-1 is a branch borehole of T3. The branch starts at 503 m. A major water loss occurred at 773 m where the coal was located. After the borehole was advanced to 779 m, the drill bit was pulled out and grouting was used to plug the leak. At 834.79 m, a large amount of fly ash and cement mixture were found to return to the surface. The drilling stopped, and grouting was performed. Coal seam #2 was found between 773 and 777 m. Yeqing limestone was found at 819 m. The strata from 834 to 902 m were soft, and the drilling rate was abnormally high at 10 m/h. The drilling cuttings returned were mixed with grouting materials. The strata became hard at 902 m, which was interpreted as the Ordovician limestone. The hole ended at 927 m.

5.4.2 Analysis of Water Inrush Passageway

Figures 5.6 and 5.7 show the conceptual site modes from interpretations of the borehole data. Some observations are presented below:

- The geologic loggings of the completed boreholes indicates that the distance between Yeqing limestone and Shanqing limestone is large. The distance encountered in each borehole is as follows: 44.1 m at T1, 42.01 m at T1-1, 47.41 m at T2, 44.25 m at T3, and 39.75 m at T3-1. The Daqing limestone is missing in all boreholes except for borehole T1-1 where Daqing limestone is found at depths between 910 and 915 m. Because Daqing limestone is encountered only in T1-1 where the distance between Daqing limestone and Ordovician limestone is 15 m.
- Based on regional geological data, the average distance between coal seam #2 and the Ordovician limestone is 155 m. In this investigation, the distance between coal seam #2 and the Ordovician limestone in borehole T1 is 149.0 m, while the distance in borehole T1-1 is 150 m. The distance between coal seam #2 and the Ordovician limestone in borehole T2 is 92 m, which is abnormally smaller. The distance between coal seam #2 and the Ordovician limestone in borehole T3-1 is 125 m. The relatively small distance in these boreholes may result from the presence of geologic structures such as faults or collapse columns.

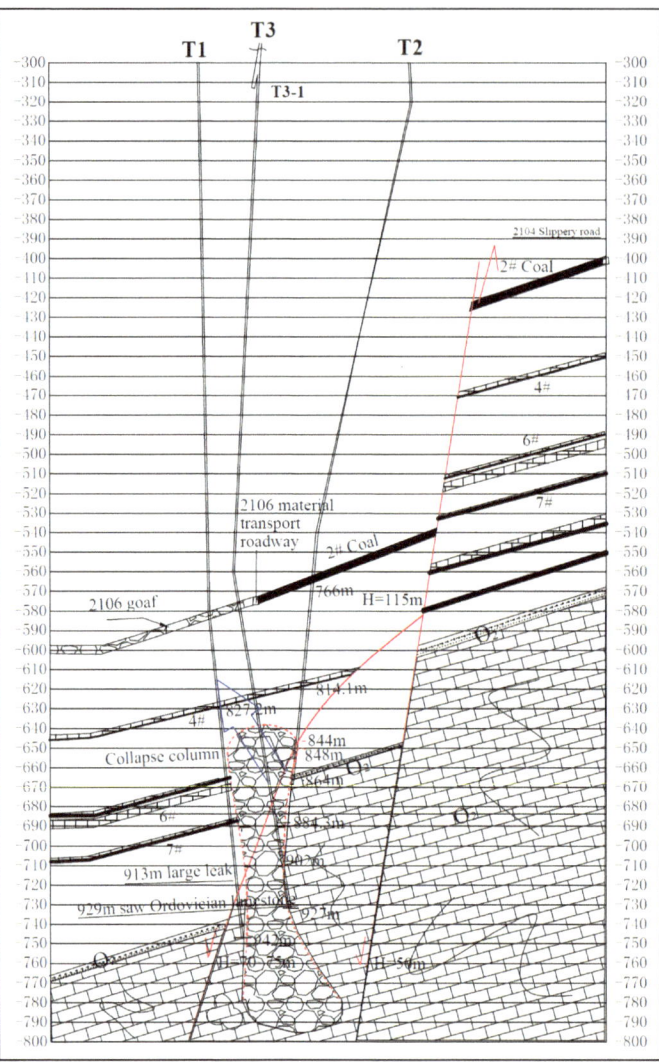

Fig. 5.6 Water inrush conceptual site model along B-B′

- The structures associated with the water inrush are normal faults and collapse columns. The normal fault has a drop of about 60–70 m and a dip of about 60–70°. The collapse column has a long axis of approximately 60–70 m and a short axis of 25–35 m. The top of the collapse column is approximately 60 m below floor of the coal seam #2 working face.
- Water-conducting fractures are induced in the coal seam floor by coal mining of the working panel. Under the persistent pressure from the underlying Ordovician limestone, the water-conducting fracture zone further developed laterally

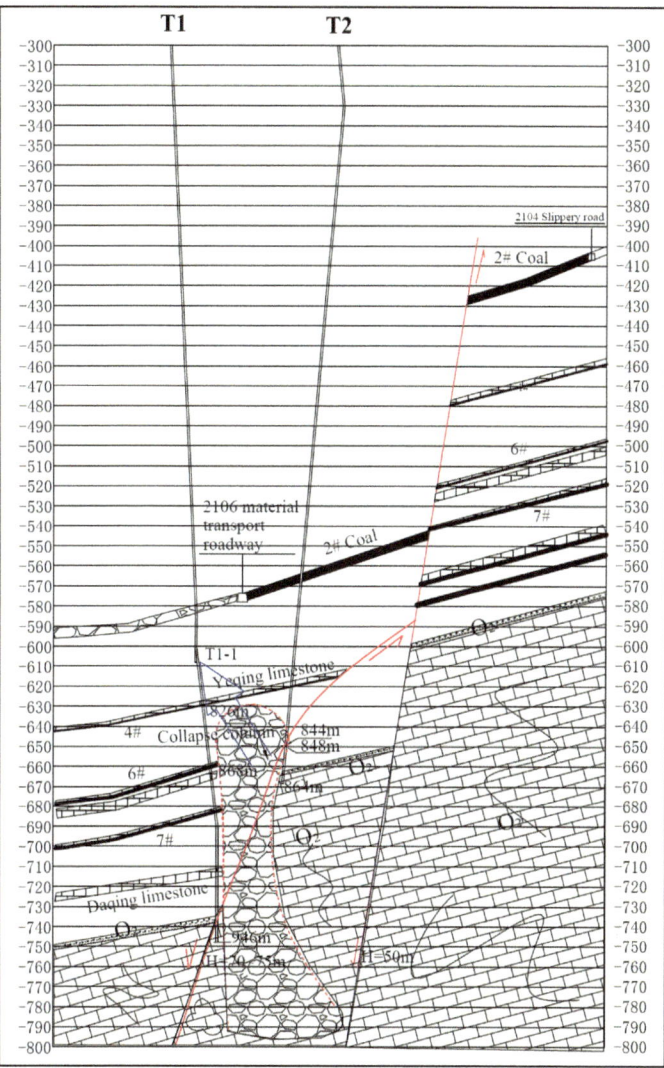

Fig. 5.7 Water inrush conceptual site model along C-C'

and vertically. When the induced fractures are connected to the collapse column and faults, the water from the Ordovician limestone forced its way through the connected passageway to the working area and caused the water inrush incident.

5.4.3 Analysis of Grouting Water Inrush Passageway

The grouting process of sealing the water inrush passageway uses an adaptive management approach, allowing onsite decisions as more data becomes available. Grouting in all boreholes is conducted under static pressures with the borehole inlet closed. The borehole inlet is tightly sealed with a borehole plugger during grouting. Well-mixed slurry was delivered from the mixing station to the borehole and injected through the borehole to the target area. If there is a large leak or drill bit drop during drilling, grouting is conducted immediately. If any abnormal condition is reported, grouting is conducted again before drilling restarts. When grouting to seal the water inrush channel, the discrete top-down grouting method is adopted. After the borehole is completed with the exploration task, grouting is conducted from top to bottom. After the pressure is noticeable, a certain time is allowed for the grout to cure. Then the borehole is advanced through the grout downward to the next grouting section. The iterative remediation process is conducted until the ending criteria are met.

During grouting, it is important to adjust the water: cement ratio of the slurry. The implementation sequence starts with the thin slurry, which is followed by the thick slurry. The slurry becomes thick by reducing the water-cement ratio to increase the specific gravity of the slurry. At the normal grouting stage, the water: cement ratio of the slurry is 1:1. If pressure builds up at the borehole inlet during final borehole grouting, a small pump volume at rates between 40 and 60 L/min is recommended to gradually reach the final pressure standard of 3 MPa and maintain for 30 min before stopping grouting. Fly ash is mixed with cement to produce the slurry to reduce the project cost, and such mixed slurry was injected in T2 borehole only. Table 5.5 summarizes the materials that are consumed in the remediation projects. A total of 32,824.28 tons of cement and fly ash were injected, of which T2 borehole had the largest grouting volume.

A water pressure test is conducted in each borehole at the end of grouting. The water intake rate or specific capacity is calculated by:

$$q = \frac{Q}{PL}$$

where

Table 5.5 Summary of grouting materials and quantities

Borehole No.	Cement (t)	Fly ash (t)	Notes
T1 (T1-1)	7363.07	0	T1-1 is the branch borehole of T1
T2	15,716.96	4595.11	No branch boreholes were constructed in this hole
T3 (T3-1)	5217.66	0	T3-1 is the branch borehole of T3
Subtotal	28,297.69	4595.11	
Total	32,824.28		

Table 5.6 Results of water pressure tests at end of grouting

Borehole No.	Pump rate (L/min)	Orifice pressure (MPa)	Length of test section (m)	Water intake rate or specific capacity (L/min mm)
T1	60	3.0	774.0	0.000258
T1-1	60	3.5	157.36	0.001089
T2	60	4.5	54.39	0.002451
T3	60	3.0	738.31	0.000271
T3-1	60	3.5	135.4	0.001266

q specific capacity, L/min mm;
Q pressure flow rate, L/min;
P pressure acting in the test section converted into hydraulic height, m;
L length of test section, m.

The calculation results of grouting boreholes are presented in Table 5.6. The specific capacity ranges from 0.000258 to 0.002451 L/min mm, which are order of magnitude less than 0.01 L/min mm. The post-grouting formations have the permeability between impermeable rock ($q < 0.001$ L/min mm) and poorly permeable rock ($0.001 \leq q \leq 0.01$ L/min mm).

5.5 Summary

Groundwater dynamics often contribute to ground subsidence in mining areas. Prevention of ground subsidence requires complete control of groundwater flow, especially during water inrush events when the groundwater levels tend to fluctuate significantly. The working face of Huangsha Mine had a water inrush with the maximum instantaneous water inrush exceeding 400 m^3/min and a rate 180 m^3/min prior to the remediation. The emergency and rescue water blocking project includes two parts: plugging of tunnels by accurate interception and sealing of water inrush passageways after pinpointing the locations. A total of 18 traditional and directional boreholes and 2 branch boreholes were completed for the concurrent investigation and remediation. The total drilling footage is 14,900.46 m. Approximately 63,123.69 m^3 of aggregates, 56,552.69 t of cement, and 4595.11 t of fly ash were injected during the grouting.

At completion of the remediation, the groundwater flow in the tunnels reduced from the initial 180 m^3/min to approximately 25.8 m^3/min, which is the normal water inflow prior to the water inrush. The groundwater levels in boreholes #7 and #10 rose to + 121.3 m and 119.74 m, respectively, which were higher than the groundwater level of + 117 m in the Ordovician limestone. After the coal mine resumed production, there was residual flow from the grouted tunnels, which suggests a water blocking rate of 100%.

Detailed drilling exploration, especially the application of directional drilling, helps understand the water inrush conceptual site model and locate the passageway for target grouting. The nature and extent of the passageway is determined from the comprehensive analysis of the abnormalities in stratigraphic positions such as the distance between Yeqing and Shanqing, the missing Daqing limestone, the distance between Daqing and the Ordovician limestone, and the distance between coal seams and the Ordovician limestone, and the drilling anomalies such as major water losses, drill bit drops, abnormal drilling speeds, and slurry channeling. Post-grouting specific capacity of the grout boreholes suggests reaching or approaching the standard of impermeable rock.